John Miller

Fetich in Theology

Doctrinalism Twin to Ritualism

John Miller

Fetich in Theology
Doctrinalism Twin to Ritualism

ISBN/EAN: 9783337253134

Printed in Europe, USA, Canada, Australia, Japan

Cover: Foto ©Lupo / pixelio.de

More available books at **www.hansebooks.com**

FETICH IN THEOLOGY;

OR

DOCTRINALISM TWIN TO RITUALISM.

BY

JOHN MILLER,

PRINCETON, N. J.

"*If we are like God, God is like us. This is the fundamental principle of all religion* . . . . *Jacobi well says :* '*We confess, therefore, to an Anthropomorphism inseparable from the conviction that man bears the image of God ; and maintain that besides this Anthropomorphism, which has always been called Theism, is nothing but* ATHEISM *or* FETICHISM.' "—Dr. Hodge, Theol. vol. i. p. 339.

NEW YORK:
DODD & MEAD, PUBLISHERS,
762 BROADWAY,
1874.

Entered according to Act of Congress, in the year 1874, by
JOHN MILLER,
In the Office of the Librarian of Congress, at Washington.

THE MIDDLETON STEREOTYPE COMPANY,
GREENPORT, LONG ISLAND.

PREFACE.

For thirty years or more the author has been busy upon a theory of Ethics. He has subjected it to every test. If it is false, he is another instance of a life wasted by error. If it is true, it justifies his absences from the pulpit; for it is of the very essence of its analysis that it sets at rest many of the questions that are dangerous in our best theology.

The author confesses that *indicia* of his special Ethics led him to entertain the scruples which this book unveils, and made disagreeable to him doctrines that have planted themselves in our common Calvinism. But these same *indicia* pointed up into the Bible, and gave him better weapons there than the novelties of an unaccepted system. If he met error by his philosophy, he would have to carry his philosophy; and that might be harder in the end than to crush the error. As a better polemic he can take the Scripture, which his philosophy suggests, and employ that base to fortify his argument. Thus he gains two things: — First, a conceded premise instead of a debated one; and second, a less suspected conclusion; for the author, having denied himself the pleasure of tying "the millstone of his philosophy around the neck" of his theology, will gain in his philosophy itself by showing in a preliminary book with what Scripture certainties his philosophy affiliates itself.

Meanwhile, the rectification which this book attempts, is the main grand purpose of his life in giving away so much of its history to ethical investigation.

JOHN MILLER.

PRINCETON, March 13th, 1874.

CONTENTS.

INTRODUCTION.................................... 15

BOOK I.

NOTHING TO WORSHIP.

CHAPTER I.
A GOD ALL FOR HIMSELF............................ 19

CHAPTER II.
A GOD WHOSE WILL IS THE GROUND OF MORAL OBLIGATION.................................... 20

CHAPTER III.
A GOD THE IDEA OF WHOM IS INNATE................ 21

CHAPTER IV.
A GOD OF WHOM VINDICATORY JUSTICE IS A PRIMORDIAL ATTRIBUTE.................................... 24

CHAPTER V.
A GOD WHOSE CHIEF END IS TO DISPLAY HIS GLORY.... 25

CHAPTER VI.
A GOD WHOSE UNIVERSE IS NOT THE BEST POSSIBLE.......26

CHAPTER VII.

A God Made Responsible for Sin by a Distinction between Preserving and Creating.............. 27

CHAPTER VIII.

Man's Helplessness not Disinclination.............. 29

CHAPTER IX.

The Faith that Saves a Man not in its Essence Moral... 33

CHAPTER X.

Rationalism an Over-use of Reason................... 37

BOOK II.

Something to Worship.

CHAPTER I.

Holiness.. 39

CHAPTER II.

God's Highest End not Himself but His Holiness.... 41

CHAPTER III.

The Ground of Moral Obligation the Excellence of Holiness.. 42

CHAPTER IV.

A God Who Does Everything for His Glory.......... 43

CHAPTER V.

The Idea of God not Innate............................ 44

CHAPTER VI.

VINDICATORY JUSTICE NOT A PRIMORDIAL ATTRIBUTE OF GOD.. 46

CHAPTER VII.

THE BEST POSSIBLE UNIVERSE........................... 47

CHAPTER VIII.

GOD NOT IMPLICATED WITH SIN, THOUGH PRESERVING PROVIDENCE BE THE SAME AS A CONTINUOUS CREATION.. 49

CHAPTER IX.

THE SINNER'S HELPLESSNESS DISINCLINATION............ 51

CHAPTER X.

SAVING FAITH IN ITS ESSENCE MORAL................... 52

CHAPTER XI.

RATIONALISM NOT TOO MUCH REASON................... 53

BOOK III.

FETICH.

CHAPTER I.

IDOLATRY A UNIVERSAL SIN............................. 57

CHAPTER II.

WHAT IS IDOLATRY?.................................... 58

CHAPTER III.

FETICH.. 60

CHAPTER IV.

THE TWO ATTRIBUTES OF FETICH THE TWO ATTRIBUTES OF THE ABOVE-DESCRIBED TEN DOCTRINAL PROPOSITIONS... 60

CHAPTER V.
HOW BEST TO MAKE THIS APPEAR....................... 62

BOOK IV.
FETICH IN PRACTICE.

CHAPTER I.
THE BIBLE A FETICH.................................... 64

CHAPTER II.
PRAYER A FETICH....................................... 65

CHAPTER III.
SERIOUSNESS A FETICH.................................. 66

CHAPTER IV.
PROFESSION A FETICH................................... 67

CHAPTER V.
ALMSGIVING A FETICH................................... 69

CHAPTER VI.
PREACHING A FETICH.................................... 70

CHAPTER VII.
FAITH A FETICH.. 77

CHAPTER VIII.
REPENTANCE A FETICH................................... 82

CHAPTER IX.

The Rationale of Fetich.................................. 83

BOOK V.
Fetich in Doctrine.
CHAPTER I.

A God all for Himself.................................... 88

§ 1. *Dr. Hodge's Statement of the Doctrine*............. 88
§ 2. *Dr. Hodge's Contradiction of his own Doctrine*..... 89
§ 3. *Texts to Sustain the Error.*....................... 91
§ 4. *Texts to Refute the Error.*........................ 94
§ 5. *Argument from Reason.*............................. 95
§ 6. *The Doctrine Fetichism.*........................... 98

CHAPTER II.

The Will of God the Ground of Moral Obligation.. 100

§ 1. *Dr. Hodge's Statement of the Doctrine.*............ 100
§ 2. *Dr. Hodge's Contradictions.*....................... 104
§ 3. *Argument from Reason.*............................. 107
§ 4. *The Error Fetich.*................................. 110

CHAPTER III.

The Idea of God Innate................................... 110

§ 1. *Dr. Hodge's Statement of the Doctrine*............. 110
§ 2. *Dr. Hodge's Contradictions.*....................... 112
§ 3. *Argument from Scripture.*.......................... 114
§ 4. *Argument from Reason.*............................. 116
§ 5. *The Idea Conceived of, a Fetich.*.................. 122

CHAPTER IV.

Vindicatory Justice as a Primordial Attribute of God... 123

§ 1. *Dr. Hodge's Statement of the Doctrine.*............ 123

	PAGE
§ 2. Dr. Hodge's Contradiction	128
§ 3. The Scriptures that Dr. Hodge Quotes	131
§ 4. Scriptures that Refute the Error	131
§ 5. Argument from Reason	132
§ 6. The Error Fetich	137

CHAPTER V.
GOD'S HIGHEST END TO DISPLAY HIS GLORY............ 143

§ 1. Dr. Hodge's Statement of the Doctrine	143
§ 2. Contradictions	144
§ 3. Scriptures	144
§ 4. Argument from Reason. Holiness the Highest End	145
§ 5. Everything Else Fetich	146

CHAPTER VI.
THIS UNIVERSE NOT THE BEST POSSIBLE............ 147

§ 1. Dr. Hodge States the Doctrine	147
§ 2. Contradictions	148
§ 3. Scriptures	149
§ 4. Reason	150
§ 5. An Orthodox Optimism	153
§ 6. The Opposite, Fetich	154

CHAPTER VII.
GOD'S PROVIDENCE NOT A CONTINUOUS CREATION, ELSE GOD THE AUTHOR OF SIN.......................... 155

§ 1. Dr. Hodge's Statement of the Doctrine	155
§ 2. Dr. Hodge's Contradictions	156
§ 3. Argument from Reason	158
§ 4. The Doctrine Fetich	164

CHAPTER VIII.
MAN'S HELPLESSNESS NOT DISINCLINATION............ 165

§ 1. Dr. Hodge's Statement of the Doctrine	165
§ 2. Dr. Hodge's Immediate Contradictions	166

Contents.   13

	PAGE
§ 3. *Scripture*	171
§ 4. *Argument from Reason*	172

CHAPTER IX.
SAVING FAITH NOT IN ITS ESSENCE MORAL.............. 180

§ 1. Dr. *Hodge's Statement of the Doctrine*............ 180
§ 2. Dr. *Hodge's Contradictions*...................... 189
§ 3. *Argument from Scripture*......................... 191
§ 4. Dr. *Hodge's Argument from Reason*................ 195
§ 5. *The True Doctrine*............................... 199
§ 6. *Anything else Fetich*............................ 204

CHAPTER X.
RATIONALISM AN OVER-USE OF REASON.................. 212

§ 1. Dr. *Hodge's Statement of his own Doctrine*........ 212
§ 2. Dr. *Hodge's Exposition of his own Statement*...... 212
§ 3. *The Doctrine False*.............................. 216
§ 4. *Its Mischiefs*................................... 219
§ 5. *The Scriptures that Dr. Hodge Quotes*............ 224
§ 6. *Injury of Dr. Hodge's Teaching to Dr. Hodge himself*.. 225
§ 7. *Argument from Reason*............................ 230
§ 8. Dr. *Hodge's Contradiction*....................... 231
§ 9. *The Whole System Fetich*......................... 232

BOOK VI.
FETICH IN ORDER.

CHAPTER I.
AS RITUALISM SHRIVELS DOCTRINE, SO DOCTRINALISM SHRIVELS THE CHURCH........................... 234

CHAPTER II.
THE TRUE IDEA OF THE CHURCH....................... 236

CHAPTER III.
Dr. Hodge's Idea of the Church 240

CHAPTER IV.
Dr. Hodge's Argument for his Idea of the Church. Its Falseness 246

CHAPTER V.
Practical Mischiefs of Dr. Hodge's Idea of the Church 254

CHAPTER VI.
Doctrinal Mischief of Dr. Hodge's Idea of the Church 257

CHAPTER VII.
On the Question, What is a True Church of God? ... 258

INTRODUCTION.

THE Rev. Charles Hodge, D. D., in his late work on "Systematic Theology," teaches, (1) that God has made everything for Himself, (2) that the will of God is the ground of moral obligation, (3) that the idea of God is innate, (4) that vindicatory justice is a primordial attribute of God, (5) that God's highest end is to display His glory, (6) that the universe is not the best possible, (7) that preserving Providence, explained as a continuous creation, is unworthy of God and makes Him responsible for sin, (8) that the helplessness of the sinner is not disinclination, (9) that saving faith is not of its essence moral, and (10) that Rationalism is an over-use of reason.

These doctrines are singularly stern. If they are true, they belong to the list of which Peter speaks (2 Peter, iii. 16), in which are "things hard to be understood, which they that are unlearned and unstable wrest unto their own destruction:" and if they are untrue, they are grossly bewildering and pernicious; like weights to a drowning man; for they load with difficulty the very points of temptation in the gospel.

Believing them to be untrue, the author knows that he would be indulged even by Dr. Hodge in trying to make that appear; and that such a polemic as Dr. Hodge should be the most tolerant of an attempt to clear orthodoxy of growths, parasitic or diseased, that might seem to be penetrating toward the life of the faith; but it will not appear so plain to anybody, and did not to the writer himself, that these growths should be traced particularly to Dr. Hodge. Why not treat them as of the Reformed belief? This was a present impulse in a scheme to notice them. But our study determined differently. Dr. Hodge, as the advanced writer, is the only authority in such things who brings them all together. Turrettin has but five or six of them. The Reformed belief has none of them; that is, they are nowhere all enforced, and they are somewhere one by one refuted. They have been forming scatteringly like crystals in a vase. Dr. Hodge has helped the process. And now, when people wake to what is going on in their religion, would it not be a sort of mock respect to appear not to see what hand has shapened them the most; and what book has made them easiest to refute by the very harmony that appears among them?

If they are God's truth (as they ought to be, to be so authoritatively set forth), no apology will save the critic from a most eccentric fate; but if they are a human error, no apology, of course, is needed. The best defence will be an industrious discussion. Dr. Hodge has so high a name, and stands so eminently among the preservers of the faith, that we are sure of our instinct when it commands us to make no excuses. If Dr. Hodge were a common man, we might less ludicrously express our regret; but as he is just who he is, every word of reserve is but that much labor lost; and every syllable of attack must keep itself in countenance only by the purest reasoning.

BOOK I.

NOTHING TO WORSHIP.

To begin, therefore :—What possibility of Worship does Dr. Hodge leave for those attributes of God that come within the sweep of his ten propositions?

Worship is from the Anglo-Saxon *weordh*, and implies that *worth* is essential to worshipfulness. If I turn a peasant into his closet, and expect him to worship, I must either give him a string of shells or some like idolatrous cheat; or depend upon his admiration. .To adore is to admire. Admiration must be intelligent, and must be able to give a reason for itself. " Ye worship ye know not what" was the crime of the Samaritans. And there is added the articulate rule—" The true worshippers shall worship the Father in spirit and in truth." We are not now asserting that Dr. Hodge's doctrines are untrue, but simply arguing that they are not worshipful.

CHAPTER I.

A GOD ALL FOR HIMSELF.

A MAN all for himself would be an intolerable nuisance. I do not assume that God's self is not so different from man's self as to be to God what man's self has no right to be to man. That we will treat hereafter. I only say that God's making all things for Himself is nothing to worship. We admire the opposite. If God is to be worshipped, we must admire him somewhere outside of this. And as this fills a wide periphery, and God's chief end makes a great figure in His temple, it is hard to see what there is outside. A stone in a furnace, filling all but the further corners, would make a man feel that there was but very little opportunity for fire; and should make him think that if he could take all out, and put in something that would burn and warm, he would be nearer the view for which the furnace was brought into being.

A God all for Himself, therefore, I do not yet aver to be a mistaken Deity; but I offer, at this preliminary stage, as having anything but a claim in Himself to adoration.

CHAPTER II.

A GOD WHOSE WILL IS THE GROUND OF MORAL OBLIGATION.

THE stride is immense. What God does everything for sweeps the universe; but what He builds morality upon reaches yet further. How can I worship without getting something intelligibly grand in both these particulars? Cut off in the direction of the first, how cruel if I should be disappointed in the latter! And yet, how can I worship anything excellent in God, if there is no such thing by a character in itself, but all is made excellent starkly by a decision of His will? (Dr. H., Theol., vol. i. p. 405.)

To say that he is the Father of our spirits, and, therefore, the norm of all possible perfection, is true, of course; but that is not the issue. A rule and a ground are very different ideas. The ground of anything being what it is, is the causal or directly efficient reason. The ground of moral obligation is that which breeds the obligation, or makes it moral; and as there is no exception in the thought, it does not apply to some forms of moral obligation, as, for example, certain positive precepts; but it applies to all morality and, of course, to the morality of the Most High.

How, therefore, is the peasant to proceed when he encounters, in worshipping, these barren thoughts? Is he to do without any conception of excellency? How is that possible? Is he to admire by the help of faith? glorifying and blessing with no power to give a reason? That has been forbidden (Jo. iv. 22). These dogmas are cruel things, when they are rooted in the very bosom of truth; coolly stated by those whom it is an eccentricity to doubt; and, yet, with consequences involved utterly alien to any adoration.

CHAPTER III.

A GOD THE IDEA OF WHOM IS INNATE.

A DOCTRINE coldly metaphysical cannot affect a peasant man one way or the other. But Dr.

Hodge hastens to explain. He says, The idea that is innate is that of Entire Supremacy (Theol., vol. i. pp. 195, 199). The peasant man will wonder,— What is that to me? A God entirely good—that I can comprehend; or a God entirely perfect; but a God simply sovereign—that may be either here or there. The Devil might be simply sovereign.

Recollect, these are absolute utterances.

(1) A God all for Himself, (2) a holiness made such at will, and (3) a Deityship whose stark idea is that He is Supreme, are weighed speeches, given out in dogmatical discourse, and all we have to go for, as yet, in reverential service.

And the latter is more a grief, because we encounter it in flying disordered from the former.

To look into the sky, and say, God is holy by a holiness made such by His will, drives me by a sort of instinct to put more body into the thought by talking of His *nature*. Dr. Hodge does this (vol. i. p. 406).

1. But will and nature are different things. They may agree. But the propositions, Holiness is made holiness by will, and, Holiness is made holiness by His nature, are not identical. A law by the will of Nero and a law from the nature of Nero might be just the opposite.

2. But if there were no difficulty of that sort, still how could we manage? Nature, as a ground, breeds the same sense of vacancy. If the nature of God is the ground of moral obligation, it is either excellent or it is not. If it is not, how can we wor-

ship? If it is, then it is excellent either by an excellence that is excellent in itself, or by an excellence that is made excellent by the nature of the most High. If it be an excellence that is excellent in itself, then the nature of God is not its ground. But if it be an excellence that is made excellent by the nature of God, then the peasant is back where he began. How can he admire God if He has no excellence that is an excellence in itself, and none that does not become such by His sovereign nature?

3. A rude heart, however, might hold on to some idea of worthiness, and push off much of this reasoning as metaphysical conceit. Dr. Hodge does not allow us to do this. "The idea of God is innate." The nest must not only be broken up, but the bees sticking to it, and desiring to rebuild where it was, must be carried off to another place. Excellence is not to be my thought at all, but *supremacy* (pp. 195, 197). The thoughts innate in my spirit are "responsibility" and "dependence" (pp. 199, 200). And, like the Darwinian scheme, I can see order in the links—that is, AUTHORITY everywhere; (1) authority as the end, (2) authority as the rule, and, now, (3) authority as the innate idea; but I do not see where the bee can rebuild her nest. Authority, whether bad or good, is not a worshipful thought; and I can clearly see its joints in a dogmatic scheme, but not its service to a habitual devotion.

CHAPTER IV.

A GOD OF WHOM VINDICATORY JUSTICE IS A PRIMORDIAL ATTRIBUTE.

I CAN worship benevolence. I can see at a glance that it admits of no exceptions. God says this. He is kind even to the unthankful and to the evil. But can I worship resentment? I turn a peasant man into his closet, and tell him God is good, and he can go off into rapturous devotion. But resentment in all spaces underneath the Supreme is wicked and forbidden. From youth to age, from savage to tutored life, from men to angels, from the lowest of the angelic host to Gabriel as he sits before the throne, revenge would be iniquity. And we may choose our own word, anger, wrath,—anything that inspires a penalty. We can make it all right enough so long as we treat it as instrumental and make it grow out of a love of right and out of regard for the stability of law, but the moment we bow down to it as a primordial trait, we are dazed immediately.

I admit that Dr. Hodge's positions are all linked together.

If right is made right simply by the will of the Almighty, then He could make that wrong in me which is right in any other being: primordially too; for we must carry things to their actual extreme, and, according to Dr. Hodge, the will of God is the ground of moral obligation.

But if resentment, which is wrong in me, is right in God, and made so by a decision of His will, I only say, The peasant man is not in His council. We are not yet challenging anything. Worship, that is the question. Gazing upon a lake of fire, and told, That is the birth of character, it is a mighty difference whether the character knows some object beyond, or is a thing by itself that must have its feast of vengeance.

And if it be this latter, then now again :—

CHAPTER V.

A GOD WHOSE CHIEF END IS TO DISPLAY HIS GLORY.

If I cannot worship God for the fact that justice is a lust of vengeance, how can I worship Him if it is all for display? I pause for no difficulties. If God's vengeance is a primordial trait it does not need to be for anything. We might hold just there. If anything be primordial it is in that very origin of it a motive to itself. If God punishes precisely as he does good, lust of punishing and lust of doing good being *pari passu*, and one just as original as the other, the inquiry as to any chief end in either is illogical in the extreme.

But that, again.

For the present, having no rest for one's feet, four doctrines having been proposed with no food for love or admiration, consider the plain man's distress if the list goes on :—A God arranging an eternal Pit, and doing it out of a lust of vengeance, but first ordaining the whole to display His power and glory.

An agony which I cannot bear, and which is to last with me forever in the flames, and which is to grow—that ghastly doctrine!—my Hell to-day being but the seed time of my torment afterward,—my Hell hereafter being but a seed time of Hell forever,— I am to accept and consider credible, not as the act of a gracious Prince in its own severities, but as the fruit of a sovereign vengeance; and not simply even that, but as an historical display, to exhibit His eternal excellency.

But mark our purpose.

It is not to discuss the doctrine. God's glory and man's glory are certainly different. Moreover, the display of God's glory will be of vast importance to the saints. Only this we are setting forth:—Here is no rapture for the worshipper. A God who leaves men in a lake of fire, and does so, as a child would say, simply to show Himself; lifting a universal frame only that way and with that utmost end, may be a God, let us try to suppose, with other and more usual traits, but must promote aside from these a creature's adoration.

CHAPTER VI.

A GOD WHOSE UNIVERSE IS NOT THE BEST POSSIBLE.

THE best possible universe is certainly better than one not the best possible. A God who creates the best possible universe is certainly to be admired, *quoad* the nature of His work, more than a God who creates a universe not the best possible. Nay, given one better, He who creates one less good may be worshipped for other features of His work, but, *quoad* that one, cannot be worshipped at all.

CHAPTER VII.

A GOD MADE RESPONSIBLE FOR SIN BY A DISTINCTION BETWEEN PRESERVING AND CREATING.

As God may be thrown out of the chambers of our worship by direct ascriptions that rob Him of all chance of adoration, so He may be inferentially defaced by unworthy arguments.

Sin being that bitter thing that God hates, to say, Unless God answers to a theory of mine, He is Himself a sinner, is a horrid risk, particularly if the theories that are to be opposed run exceedingly close, the one to the other.

God creates me and foreknows me and predicts me, yea, predestines me; He upholds me, and concurs with me, and, if He withdraw His hand, I sink into nothing in a moment. This is one theory; and gives God, so Dr. Hodge declares, no painful responsibility for sin. The other theory is, that God creates me and continues to create by a continual emanation of His power, just as it was in the beginning. Now, I do not argue. This is not the time for it. I make no choice. I have as good a right to one theory as the other. But Dr. Hodge, in risking everything upon a rational conceit; tying the millstone of his philosophy about the neck of his belief in God (Theol., vol. ii. p. 73); coloring a shadowy test, and saying, this theory makes God good, and this theory makes Him the fountain of our trespass, is giving the neophyte a new push out of the temple. Let us look at these things again. God creates me. Then he knows distinctly all that I am to be. He decrees me, and has me mapped before Him in all that I am to do. He upholds me, concurring with me in my work, and flowing into me as the only means to keep me from annihilation; and Dr. Hodge adopts all this as utterly consistent with God's not being responsible for trespass. But if I say, All this is unnecessary; the truth is far more simple; brute atoms are but the Ahriman of the East, utterly superstitious as shields to the Majesty on high, Dr. Hodge breaks out with all his thunders.

I say, This is not safe. To launch the bolt of accusation against the God that made us; to say my theory is true, or God is a liar; Christ did not make wine at Cana of Galilee, or Christ is no Christ for me; God did not touch slavery on the top of Sinai, or God is no God: all these are kindred hazardousnesses.

Being not being unless God is in it to hold it up, and being not being unless God continues to create, are to me *in se* so similar, that to say, that in one case God is healthily aloof, and in the other case visibly responsible for our trespass, is in all views singularly venturous, and the more so with a man who believes that souls are freshly created (vol. ii. p. 70), that is, the infants of our species, not being traduced from Adam, but directly created in their sins by the Almighty.

Our charges, therefore, are, first, that he wounds both theories by wounding either; second, that he cannot keep holiness out of one without keeping it out of the other; and thirdly, that he is standing on a Rationalistic foot (strange as that seems for Dr. Hodge), because he is confiding the worship of the Blessed to the keeping of a scientific investigation.

CHAPTER VIII.

MAN'S HELPLESSNESS NOT DISINCLINATION.

ENGAGED heretofore in the region of Natural Religion we come now nearer to the gospel.

If God has made everything for Himself, and even virtue in its original distinction is manufactured by His will; if the idea of Him is innate, and that idea is the idea of a Naked Sovereignty; if vindicatory justice is primordial like grace, and the same Sovereign God delights in it for His personal display, we might hope that, when the doors of mercy were opened, and we should get nearer to the cross, the sky would lighten. And yet Doctrinalism, with the same harsh airs, comes in to the region of salvation.

We are helpless.

Now, of course, the Scriptures teach that we are helpless. It is a doctrine that all confess.

We cannot exaggerate the entireness of our helplessness, for it is a helplessness like death. We cannot wake up. And if we cannot wake up, we cannot hear or think or feel the doctrines of redemption. We can be waked up only by the Spirit.

But the question is, Is not disinclination the cause of it?

Dr. Hodge says, No.

I have been preaching that it is. The sinner is cold and dead, but why cavil? he is wilfully so. And I strip all hardship out of the case, for I show that his whole abandonment of God is from the state of his affection.

I do not preach less helplessness, but more wilfulness; and I do not underrate at all the dead condition of the sinner because I take off the blame from God and put on the blame upon the lost man's wickedness.

I exalt God, too, more, after that, in the preaching of the gospel.

I call to a prisoner. The jail is burning. I entreat him to come out. He refuses. I pity him less, because his doom is wilful. But I search the ashes, and the wretch was chained! My cry trifled with him.

Dr. Hodge seems quite impervious to such practical ideas.

He says (vol. ii. p. 70): We are all originated separately, traductively in our bodies but separately in our spirits. Created in this way apart, we owe our relation together to a covenant with Adam. Born each man by himself, it has pleased the Almighty that we should be born wicked. And born wicked, the inference is that I came into the world helpless, and now, even after the offer of the gospel, that I have never a chance to use my inclination, or in any single instance to have a choice whether I will not repent and become a better man.

Can the peasant take this and turn it into the coefficiencies of worship?

I might say, You are lost; you deserve to be; you refuse the opposite; you will not come to Christ that you might have life; and I do not nullify the fact that you are dead, but I do the thought that you can have a right to cavil. You are lost because you choose to be. But if I weld all that other chain: a soul created and not born; wicked out of the very hand of God; helpless out of the very nature of its wickedness; helpless like a dull log, and not because it refuses to be delivered; with no chance, therefore, in all its history to say whether it will have anything better or not, I do make grace a cheat, and the sounding alarums of the gospel a mere tom-tom for the misleading of my spirit.

If I preach:—True you are helpless, but not helpless in any ordinary sense: God has truly provided pardon,—generously offers it,—would mock you were you in all sense helpless: you are in one sense helpless, and in such sense helpless that you will never live through all eternity without the grace of Emmanuel; and yet, your helplessness is being unwilling. If I say, the drunkard is helpless, and he is chained to his cups, but yet his helplessness is helplessness of will and desperate appetite, you cannot complain of this; for we do not allow a culprit to complain of misery if he has been repeatedly helped up, and brutally unwilling to maintain his liberty.

Yet Dr. Hodge declares that all this is utterly unsound. Helplessness (under which, remember, God creates us, and creates us *de novo*, so that we have never had a chance for a nature different) is so absolute as not to be imputed to the will, and not to be recovered from whatever the effort of the will on the part of the sinner. Plain men are horrified! No man pretends that reason can lay hold of this. God creates me, and creates me wicked, and sends me Christ, and offers me salvation, and begs me to accept it, and brands me if I refuse, and yet all the time had created me unable, and that inability not consisting in an unwillingness to try, but in the blank impossible of a withheld salvation. Now we take nothing for granted. We do not denounce this as heresy. We are not at that stage of our work. But we do declare,—Here is nothing to be worshipped; and a plain man must be turned still out of another chamber of the Temple, and must find some other *memorabilia* of God to warm him and to bridge for him afresh this other surd point in the statement of His character.

CHAPTER IX.

THE FAITH THAT SAVES A MAN NOT IN ITS ESSENCE MORAL.

When I hear that baptism saves a man, I shudder. That felicity for eternal years and ransom from the horrors of the Pit are to follow a little oxygen and hydrogen applied by the fingers of a priest in an earthly instant, I shiver at being obliged to believe anything so little. No mention of a reason can give it dignity, and no comparison with the Apple in the Garden (Gen. ii. 16), or with the clay on the eyes of the blind (Jo. ix. 6), can rob it of its look, or give it the least especial dignity. Christ has died, and I know tremendous odds have been paid for our redemption, but still, if any conditions are affixed, to be worshipful they must be proportioned to their purpose. God could convert me at His pleasure. He could visit me when asleep. I could retire accursed, and wake glorified. He could convert me by the Ten Commandments. Christ, having once suffered, He could apply life as He pleased. He could use any system of truth, or use none, or save me without knowing of His sacrifice, or by psalm-singing, or circumcision, or wearing a particular coat, just as He might choose to do. That He should limit grace to those that hear of Himself, I can see reasons for, but no imperative need such as should force it to be the system.

I cannot argue, it must be confessed, as strongly as before. God *might* demand faith, and make it all of the intellect. Still, if I am to worship—and the ways of God are all that I can think of for His praise—I see a difference between saving a man for his intellect, and saving him through piety of heart.

There are two schools, therefore, that might emerge on this important question. All believe that we are saved by truth. The poor heathen, as one of the terriblenesses of Scripture, are supposed to perish. We have five senses. Through some of those channels must come a story of the Cross, or we must necessarily be lost. This is hard to be received. But all, after patient toil, begin to see reasons why this might wisely have been decreed. But when men go further, and say that there is a difference among souls that hear the gospel,—that some men hear it savingly, and some men hear it in such a way that it increases their condemnation, we naturally turn with eager interest to that distinction, and naturally ask that it shall wear a dignity at least commensurate with the mighty difference.

Now, what discrepance would be more complete than one of holiness? What curses us is sin; what lifts us is a return to holiness. Looked at on the side of God, our conversion is a new birth. Coincidently, on the side of man it is an act of faith. Why not judge of the two things together? The new birth is a moral change. Why not faith a moral act? A lost man may be ever so much intelligent. He may think like a sage. He may speak like a seraph. He may expound doctrines like the apostles of God. But he is not new-born, because he has his old sinful nature. Why not speak that way about faith? I may have all faith, so that I can remove mountains; I may believe articulately and in every imaginable way; I may understand and distinguish and trust; I may be convinced, and determined and resolute in upholding religion; I may believe in every possible way, except one, and not be a Christian; and why not make that one way to be the obverse of regeneration—i. e., a moral faith? Why not agree on that as the Church's teaching? A man owns a painting, and understands it perfectly. He gave a large sum for it. He understands its shape and colors, and is a fine judge of distances. He comprehends perspective, and can descant knowingly upon the perfection of its parts; and yet he has no faith in it. He tells you he does not admire it. Visitors batter at his gates, and other men enjoy the picture; but it gives him no pleasure. What is the difficulty? He has no faith. He has faith,—a plenty of it,—but it is of the mere intellect kind. He has no æsthetic faith. Let me touch him, and let my touch be God-like in its efficacy, and it shall change him into one thing. He shall have *taste*. The eyes of his understanding shall be enlightened, and the light shall be that one thing—beauty, and it shall flash over all the rest, and his intellectual faith shall be pervaded with an æsthetic character. The image is complete. Here is a man possesses a Saviour. He is hung up in his house. He is visibly set forth crucified before him. His next door neighbor is a Christian, but he understands the picture less well than this friend who is lost. The saved man takes lesson from the lost man in the truth of religion. What is the difference? Precisely as in the other case—a want of *taste*. Let me touch him with my finger, as with Almighty grace, and what is the result? Simply a moral one. He has all the rest. As with the man with the picture, he understands all the facts. What he needs is light; and he has all sorts of light except one sort, and that sort is moral light, and that light illuminates every other, and unites the doctrines that are concerned in this man's salvation.

Now I ask whether distinctly excluding this from the essence of faith (see vol. iii. pp. 41, 72), and making faith consist in something that precedes this, and simply leads to it, does not in the first place let down the bars, and give men no certain test for supposing themselves religious; does not in the second place, arrest conversion; does not, in the third place, fill the church with hypocrites; does not, in the fourth place, keep them so, faith being an unmoral faith and endangering that there be only an unmoral sanctification; and does not, in the fifth place, dampen the parents' work in a moral, careful training toward the faith which is to be the soul's salvation.

We have gone a little deeper than we need, for we only meant to speak to this point, that faith, as a moral sight reaching down to all the distance that is regenerate, is the true counterpart of birth, and the *worshipful* condition of the soul's salvation.

To call it not moral is to exclude all that it is distinctively; for though it must take hold intelligently of Christ, yet it did that before, in all but moral respects, when there was no saving hold upon His redeeming excellency.

CHAPTER X.

RATIONALISM AN OVER-USE OF REASON.

It is odd that a mind that would keep Faith within the periphery of Reason should exclude Reason so from the domain of Faith.

We barely glance at this.

We will recur to it again where we can do more justice to Dr. Hodge. He does not introduce this point as directly as the others. In fact, he exalts Reason sometimes till one shrinks from following him. For the present we intend only this picture: —A man turned into his closet and told to worship God: when he only bows or only mutters, told,—No, it must be an intelligent admiration; when he asks, For what? told,—for His great virtue; when he asks, What is His chief virtue? told,—His making everything for Himself; when he asks, Why is that a great virtue? told,—Because that is His will, and the will of God is the ground of moral obligation; when declaring that he cannot see that, told that he does, for the idea of God, and particularly of this responsibility to God, is innate; when asking for some other virtue, told,—Vindicatory Justice; and when asking, Why that then is thought so wrong in men? told,—Such is the will of God; when asking, What greatest end God's will is driving towards with such expensive methods as His vengeance, told,—To the display of His own glory; when asking if that leads to man's best good, told,—Not necessarily; when roused against this, and complaining that God must then be responsible for the wicked, told,—O no; He only creates and foreknows and upholds and predestines and concurs with them in wickedness, and that this, by virtue of a certain secondary or causal subsistence of men, is vastly different from a continuous creation; when asking if there is no way out of this entangled lot, told,—Yes; but that man is helpless to find it, and that this helplessness does not consist in disinclination; when asking, Why not? seeing that sin itself consists in disinclination, and that the first step in a return must consist in a better will, told,—No; the first step in return consists in Faith, and Faith does not consist in what is in its essence moral; and when saying that this is a total confusion of Reason, and something that the mind cannot steady itself under, or plough its way out of, with any argument, told that it need not, that Reason must be a sort of outlaw in religion, for that the over-use of Reason is a skeptic Rationalism.

BOOK II.

SOMETHING TO WORSHIP.

STILL shunning direct argument, it would be intensely interesting if we could find a common law for these ten points of Dr. Hodge, and before treating them one by one could detect a practical mistake that could account for all of them.

CHAPTER I.

HOLINESS.

THIS is a first class quality, and hundreds of pages in Theology, without any great mention of it, must be objects of suspicion. In the hands of Dr. Hodge all that is written must have the presumption of being orthodox, and the very sound of the ten doctrines that we have marked will be orthodox in Presbyterian ears.

But suppose we can create a distinction. Suppose we dissect away these ten propositions. Suppose we can show that Calvinism is complete without them. Suppose they are excrescences. Remember, they have never been together before in any Calvinistic book. Suppose the Church seems growing in them, and getting crusted by them. And suppose (what is now the task) we can generalize them and show the seminal mistake, and that that mistake is nothing less than a dropping out of HOLINESS, cannot we abate the prejudice that would unify all the work of Dr. Hodge, and, with a cooler eye, fix only upon the genuine part of it?

To do this let us define *Holiness*.

Holiness, by way of preface let it be observed, is either, first, a quality—in which sense we speak of the holiness of a certain act; or, second, the act itself—as when we assume as holiness an act of love; or, third, a character. It is in the second meaning of the three that we shall be considering Holiness in this present chapter.

In this sense it consists of two things—Benevolence, and the Love of the Quality, which was the earlier meaning.

This appears in the Two Commandments: " Thou shalt love the Lord thy God;" that means, —Thou shalt love His Holiness. For sovereignty, and for potency and immensity and sagacity, and whatever things are not moral, we are not obliged to love Him.

But God is like man, and God's holiness and man's holiness are alike, because one is in the Other's image. God's Holiness, therefore, is (1), a love to the quality of right, and (2), a love for the welfare of all His creatures.

God's Holiness is His Highest motive. Being His highest motive, God's Holiness is His imperative law. Being His imperative law, it is fair to distinguish and to say,—God's love for the welfare of all His creatures is less, *quoad* the obligation of motive, than His love to holiness.

Here Dr. Hodge has two suspiciousnesses:— first, that these two holinesses take no part in the ten propositions that are before us; and, second, that the two thousand pages of his work put Love outside of Holiness, give scarce three pages to the latter, and give not thirty pages to the whole ungeneralized list of Jehovah's excellencies.

Is this a mark of decay in doctrinal theology?

The Bible fairly thrills with ascriptions to Holiness.

Let us look, therefore, again at the ten points;— not now to show that they have Nothing to Worship, but what they need to give that; specifically, what is the crook in them, and what that is which, being supplied, turns all straight, and gives flesh on these very bones for the body of our Theology.

CHAPTER II.

GOD'S HIGHEST END NOT HIMSELF, BUT HIS HOLINESS.

HOLINESS being a love of right, and a love of others' welfare, God's Holiness consists in these; and a love of right being the more imperative of the two, and the most imperative of all the motives that can be conceived, God's Holiness is His Highest Motive. This softens the proposition at once. A man goes into his closet, and has something to adore. Nay, he changes to the highest admiration. A God all for Himself he looks hard at. It helps him never a whit. But a God all for His Holiness rises at once to the very highest reaches of his praise. He understands that perfectly. And instead of a stone of absolute offence, he turns round this marble in the building, and it fits confest in its place upon the wall.

CHAPTER III.

THE GROUND OF MORAL OBLIGATION THE EXCELLENCE OF HOLINESS.

THAT the Ground of Moral Obligation is the Will of God must all bristle with mistakes. Can He make right wrong? Can He make right so that there is none till He chooses to make it, and that there is none as a moral obligation to Himself; and that can be His glory till He wills to have it so, and till He stamps it as right, and puts it in His character?

We will not forestall this argument; but simply show how difficulties vanish when we submit another ground, and bring in again for that the excellence of Holiness.

If I tell a plain man to praise God because He makes law and says what shall be right, he stares at me. But if I tell him, God honors law and loves right, there is a ground of adoration at once. He goes cheerfully into his closet, and admires the Prince who binds him with a law made right by its intrinsic excellence.

CHAPTER IV.

A GOD WHO DOES EVERYTHING FOR HIS GLORY.

A GOD who does everything to *display* His glory, leaves me evidently to ask,—what is His motive in that? Dr. Hodge has hardly enounced his proposition before he is borne helplessly that way, and begins naively to tell us *what good God does* by exhibiting Himself to His created universe. But the good He does mars the picture, because it instantly suggests that the *Display* cannot be the motive in itself. The *Display* of excellence cannot be God's Highest End, because it leaves room to cut in behind it, in that we can immediately ask,— What is His chief end in that? Moreover, for a Great God it is an irreverent notion that His Chief End is such a thing as show, and that His Whole Design is one that terminates, i. e. has its end, upon the creature.

Now bring in Holiness. It comes in like magic to smooth everything that looks like difficulty. Solomon calls it Wisdom. He says, "I was set up from everlasting" (Prov. viii. 23). He speaks of this very thing, the planning of the Universe. He says, "Jehovah got possession of me as the beginning of His way" (*ibid.* v. 22). He says, "I was by His side a builder" (v. 30).* And leads us to infer as the grand doctrine of his book, that the

---

* See Author's Comm. on Prov., *in loco.*

chief end of God was "*Wisdom*," or, as he had defined it, *Holiness* (Prov. i. 2, 3, *See Comm.*).

Now bring in the expressions of the Bible. We are told that the chief end of God is His "*glory*." But when we come to consider this Hebrew word, it has nothing to do with His display. If the idea of showing Himself is broached, it is only for an inferior end. Where His *chief* end is noticed, it is declared to be His Glory; and where glory comes to be expounded, it is found to mean simply *weight*. It is the *weight* of God, or His intrinsic excellence, that is found to be the highest aim of His activity.

To say, then, that the highest aim of God is to *show* Himself, confounds the worshipper. To say that the highest aim of God is to *be* Himself, relieves all, at once. Holiness immediately comes in upon the stage. It is Holiness that has been kept out. Glory in the sense of *weight*, that is, Holiness as God's highest excellence, comes in at once upon the scene, and takes its place legitimately and as His highest object.

CHAPTER V.

THE IDEA OF GOD NOT INNATE.

Nor should we allow the Holiness of God to burrow out of sight as we do by admitting that "the Idea of God is Innate."

We postpone argument in chief. We are not showing that these propositions are incorrect. We have no right to take for granted that they are the propositions of Dr. Hodge. But, postponing the test of them to the appropriate portion of the book, we are only showing now the absence of the great element of Holiness.

Holiness is not an innate idea. It has three senses. In one sense it means a quality. This quality is not an innate idea; we detect it as we do beauty, by direct conscious perception. It is a quality of two emotions! These are holinesses, as by the second meaning of the word; and these are not, of course, innate, but are detected, like the fragrance of the rose, as absolute perceptions. The third meaning is a character. This is neither innate nor conscious, but an observed law, not inspected in itself, but known of through a manifold experience.

A peasant worshipping an innate idea is a befogged and unsettled imbecile. But bring holiness in, and remind him of what right act is in his own nature, and make that infinite in God, and he has something tangible at once. Our generalization is still complete. The ten points nakedly give nothing to worship. But alter each by bringing holiness into the account, and it touches all as with a talisman. The grim face drops off, and underneath, as from a hideous mask, comes a sight for the very tenderest admiration.

CHAPTER VI.

VINDICATORY JUSTICE NOT A PRIMORDIAL ATTRIBUTE OF GOD.

WHY should it be? If Holiness is God's highest motive, then nothing is primordial except its two only exercises. In fact, one of these is not in chief, for love to creatures is not equal in its binding force to love to the principle of Holiness. Love to the principle of Holiness is final and supreme.

If I say, God loves the principle of Holiness, and, in order to advance it, He punishes, because pain and curse are natural instruments for discouraging sin, I place God distinctly in analogy with other beings. I do not subject law to mere happiness, because I do not make benevolence the ruling trait. I bring in that which is really august and final. I do not exalt vengeance, but put something altogether above it, and strip it intelligently of a leading place. I make all Scripture at once consistent. If God says, He pities me, I know how that is consistent with His vengeance. If He says, He would have all men to repent, I see at once that vengeance is not primary; I see at once that benevolence, which is itself not chief, nevertheless is nearer to the head than the divine resentment. Resentment is not a trait *re ipsa* at all. The peasant man may look at God exactly as at his fellow. The primal attribute of God is Holiness. The primal exercise of Holiness is love to its own quality. A way to develop Holiness is to punish. Vindicatory Justice means a fidelity to punish. The efficacy of punishment lies in the very constitution of Nature. Vindicatory Justice is, therefore, as of the very constitution of all. But, spite of this claim to naturalness, it is a mere secondary trait.* So the plain man can worship it. And thus again he is extricated from the dark. How can I love God for revenging, when I am to hate it in man? Answer: He does not revenge. He is the world's Magistrate as of the analogy of His creatures. He hates to take vengeance (Lam. iii. 33, 34; Ezek. xxxiii. 11). His primary desire is holiness. Punishment is its needful aid; not always for the subject of punishment himself, but for the greater universe. A penal law is a contributor, a gloomy second, to the Holiness which is His aim.

CHAPTER VII.

THE BEST POSSIBLE UNIVERSE.

HENCE this is the best possible universe.

If holiness is God's highest end, then His own holiness must be; for, that being His supreme desire, it is inevitable that He should indulge it. He being able to plan *de novo*, and not having an Ahriman to contend with, would certainly form the scheme of a universe that would be the very holiest, whether it be happy or no.

I admit that His own holiness precedes. So, of course, He would love His own holiness more than that of any of His creation. But, holiness consisting of two emotions, I cannot conceive that the holiness of God should dwarf the holiness of any of His creatures. His own holiness, therefore, not interfering with the rest, we may treat His own as out of the account, and say that, *quoad* the creatures He has made, His highest end is their utmost holiness.

Of two things, therefore, one:— Either God's highest end is not actually reached, or else, from the very nature of His holiness, this universe is the very holiest that could be made.

But, further; benevolence is the other emotion; inferior, no doubt, and greedily sacrificed if the other demand it; but, specifically, how could it demand it? Conceive the solecism! That this universe is the holiest possible, is a queer reason certainly why it cannot be the happiest possible; and, therefore, it can, and is, both holiest and happiest, unless a stop is found in the ruling emotion.

We claim both therefore. Difficulties we can treat afterward.

Holiness is our stand-by. And if that is all right, and it clears away another difficulty, so that the peasant can build where there was a pit, and breathe where there was a mephitic vapor, we count that, evidence in the very fact, for it illumines with its mighty difference. A God whose whole work is the best, and a God whose whole work is not the best, are very discrepant Jehovahs; and the closets where the two are met, must glow with different fires, and shelter different ranks of astonished worshippers.

CHAPTER VIII.

GOD NOT IMPLICATED WITH SIN, THOUGH PRESERVING PROVIDENCE BE THE SAME AS A CONTINUOUS CREATION.

WITH all the rough temper of previous propositions turned away, we come with more cheerfulness to face the difficulty of God's responsibility for sin.

God *is* responsible for sin very radically, and in ways that are innocent: for remember, He created the universe, and sin is in it.

Dr. Hodge drives this to an extreme; for he makes God create the infant, and create him wicked, and not traductively in any natural sort from the souls of our first parents.

To be responsible in ways not innocent must arise from one of two things; either, first, that God sins our sins, or second, that He sins by connexion with us who sin them. Holiness appears again as our ally to grapple the difficulties of this new dilemma.

What is holiness? It is an emotion of benevolence, or else it is an emotion of love to right. What is sin? It must be a want of benevolence, or a want of love to the quality of right. The first two are emotions; the last two are the want of them. It is clear God cannot commit the sin itself, because, however responsible for it, it is the act of a certain state of a heart's emotion. Whatever His relation to that heart, it is the heart's act, and not the act of the Almighty.

3

But He may be responsible for it!
Yet how?
We have seen that He *is* responsible for it, and that in the most important ways.

To be responsible for it *sinfully* He must be so in one of two manners,—either by failing in benevolence, or by failing in love to the principle of holiness.

Let it be understood positively, God cannot sin, just as man cannot sin, except in these sheer particulars,—a want of love to the welfare of others, or a want of love to the principle of holiness.

Now what if both objects are promoted by ordaining sin?

I pretend to no full defence. I do not claim that my theory sweeps the difficulty. I only say,— It softens it. Having found a cause for the best possible universe, I only say that we may be enraptured by this good Jehovah, and trust, more than Dr. Hodge can, that out of the horrors of the lost His holiness can maintain its objects.

But what if it cannot! What if other principles must be brought in to the solution: one thing is certain,—making God holy in the way man is holy, and bringing down the trait to the level of our intelligent emotions, makes the distinction that Dr. Hodge drew still more conspicuously unsafe. If holiness is these two emotions, it is contradicted by a preserving Providence just as much as by a continuous creation. To say the least, it is a very rash distinction. In the light of a human holiness to say that God is perfect if He ordains and concurs, and is a sinner if He continuously creates, is to launch a bolt, of which the deeper men think the more it will wound their respect for the Almighty.

CHAPTER IX.

THE SINNER'S HELPLESSNESS DISINCLINATION.

IF holiness be a love for others and a love for the principle of moral right, sinfulness, which is the opposite of these, must be the same numerically as disinclination. If sinfulness be the same as disinclination, helplessness, which is the same as sinfulness (and I mean by that numerically—not that the words do not have a different aspect), must be the same numerically as disinclination. In other words, you must find some other account of a sinner's helplessness than that it is of the very essence of his sinfulness, or else you must admit his disinclination as of the very essence of his helplessness also.

Please notice how in this first view Dr. Hodge has pushed aside the moral feature from his system.

But again, restoring holiness as not finding its ground by the will of the Almighty, and not belonging to a God in idea innate, we are warmed by a human conscience, and come into the region of familiar right. Thus judged, Dr. Hodge's idea of helplessness becomes impossible.

The idea of helplessness at all is difficult; but make it moral, and bring it within the region of man's disinclination to repent, and it mingles with a thousand facts of the helplessness of crime in this world. But make it spiritual, as Dr. Hodge would see fit to call it, and make it of a nature of spiritism that is mystic and kin to the innate, and then bring down upon it a plain definition of holiness that is in the region of our race ; and then, inviting after creating helpless, and promising where there is no strength to accept, and cursing for this where the helplessness is not disinclination, appears in its true deformity. It is an outrage upon the holiness of God; and, therefore, let this distinctly appear,—that it has been a leaving out of holiness that has enabled Dr. Hodge even to broach his ghostly proposition.

CHAPTER X.

SAVING FAITH IN ITS ESSENCE MORAL.

HERE the moral is *pro forma* left out.

We beg the reader to mark this special generalization.

Ten propositions have been found singularly unfavorable to worship. We have sought some character for them. We shall seek hereafter (p. 60) some origin in a failing of the church. The character we have found is in an omission. The omission is of the attribute of holiness. And in this present instance among the ten the omission is most direct. We have reached the great act of *faith;* and instead of making it answer to the great fact of regeneration, the interests of sin have succeeded in keeping it distinct. It is made something that logically precedes repentance; and, instead of counting it the dawning light that shines brighter and brighter to the perfect day, it is made intelligent first and moral afterward. Dr. Hodge would deny that sanctifiedness was but a higher and higher exercise of faith, or else he must affirm that sanctifiedness is not moral, or else he must deny that the first act of faith is like (only lower) all the other acts of faith that come after and spring out of its loins.

Hence we might be sure there would be hypocrisy in the modern church. These dogmas both follow and produce it. A sinner let in upon Christ upon a ticket not of its essence moral, stays in upon no better, and it fills the church with earthliness. All the propositions conspire. A God quite for Himself, a morals manufactured by a will, a Deity mystically innate, a Judge removed from us by a passion for revenge, doing all things for display, and doing nothing for the best,—all sap the very foundations of our godliness, break in upon the Temple of our praise, and give us no certain thought about either self or a Creator.

CHAPTER XI.

RATIONALISM NOT TOO MUCH REASON.

WE would not wonder, therefore, that such a system should be intolerant of reason.

Reason is universal in man. It judges everything. It takes cognizance of the spirit. It is the instrument of life. It must bring us to heaven.

Reason is but the mind of man as it takes cognizance of truth. Truth is of all sorts,—as for example the beauty of a cloud, or the excellence of holiness. Too much reason is like too much grace or too much piety; and the very name of it sells us to the wicked. Philosophers rejoice when with their wicked will they betray the divine into discoursing against reason, and using that word Rationalism as tantamount to something too keenly rational.

Reason is a revelation of God. It is an earlier revelation than the Bible. The Bible has been built on reason. It could be no revelation without reason. It is made out as revelation only by reason; and a reason sanctified is only a restored reason, and a reason faithful to its trust and risen to its highest reasonableness.

It is fearfully dangerous to decry reason before unregenerate men.

And now, as before, reason is cast out as the genuine friend of holiness. If God be innate we get our idea less reasonably. If He be for Himself, and loves display, and finds it congenial to take vengeance, and (without all that tedious list) is and does all that has been proclaimed, reason is less necessary. Holiness restores reason;—that is our interesting idea. Holiness restored to God brings Him within the study of our reason. And if that Holiness be the same as man's we instantly begin to worship it. We pass over the bridge in our prayers, and lay hold of something intelligibly excellent. Reason is our friend before the throne. And if holiness is benevolence and the love of virtue, we can see how those plain things could be understood ; and how much it would be for the interest of sin to keep reason hid, and to have it defamed and suspected among sinners.

BOOK III.

FETICH.

IT is time now that we trace the origin of these errors of the orthodox. We have shown their character, and shown that it consists in an absence of holiness.

A physician does two things :—He finds the symptoms of a disease, and then the source of it, i. e., the stone or the ulcer to which it can be traced.

What Dr. Hodge says is symptomatic of his Church ;—not necessarily not helped on by him, but growing that way in the orthodoxy in which he has been upreared ; and growing that way by just such men as he,—just as a tree grows by its loftiest and topmost boughs.

Where then is the root of mischief?

If aught be wrong, these things are horridly and growingly wrong ; and where will they end ? for they feed upon the best things in the gospel.

The humility that doubts man's work ought not to be unorthodox ; and why not take man's orthodoxy to pieces ? Mussulmans and Boodhist quietists we have given up ; Papists and superstitious Greeks ; Ritualists and Rationalists ; Arminians and New England divines ; but have sketched, like Mussulmans, our own grim-faced King, and called Him Allah.

Have we not forgotten one text :—"FEAR GOD AND KEEP HIS COMMANDMENTS, FOR THIS IS THE WHOLE OF MAN?"* (Ec. xii. 13). And may it not be true that God chooses to expose all this; to show that doctrine is no defence ; to proclaim that orthodoxy can decay with the blush upon its cheek ; to expound how orthodoxy can become a snare ; and to declare that as the beautiful orders of religion can be turned into a curse, so can its doctrines ; and that RITUALISTS can be joined by DOCTRINALISTS on a common ground of debauching piety ?

CHAPTER I.

IDOLATRY A UNIVERSAL SIN.

IF the Arch Fiend intended to seduce us, would he not conceal the motions of superstition, long after the taint had entered? Idolatry is a human appetite, native and never absent from a people. It may be disguised. So it was among the Israelites ; and so it was among the Fathers of the Forest. They began monkery, but they did not know it ; and they began the *opus operatum*, but would have shuddered to find it out. That we have no dream of being beset by idols is, empirically, the mark that should awaken our greediest suspicions. We have no rites, and few symbols, and but two sacra-

* " Duty " *is in Italics.*

3*

ments, and no pictures, and the very fewest forms that are possible in any system of worship. We have no priest, and no shrift, and no keys, and no powers, such as have been understood among the cult-systems of men. Paul might say,—' I perceive you seem to yourselves to be altogether unsuperstitious.' What has become of superstition? Is it not general? Does not Paul say (Gal. v. 19, 20),— "The works of the flesh are manifest which are these—Idolatry, witchcraft," etc.? Does he not sum up declension by talking of "[changing] the glory of the uncorruptible God," and of "[worshipping and serving] the creature more than the Creator" (Rom. i. 23, 25)? And, unless men have changed prodigiously since that day, where is idolatry with us? Or where with us specially orthodox (for we all claim that) is that special superstitiousness which Paul would say is imbedded in every one; which we see breaking out in Ritualists and in the Papal Church, and which we set no guard on, and have no special thought for, among ourselves?

CHAPTER II.

WHAT IS IDOLATRY?

IDOLATRY is worshipping something else than the genuine Deity.

Idolatry is the worship of idols; and there are men that might insist upon that as the naked idea. But as the worship of idols is always with an interior sense, and few worship the image without some regard to a Deity beside, it is safe enough to suppose that idolatry, as the wrong cult,—that is, as worshipping something not a Divinity,—will be sufficiently accepted as the meaning of the evil.

Now what is a wrong object? Eminently a SOMETHING NOT HOLY. If we might worship God as powerful, or if we might worship God as wise, if we might worship God as great, or, as the Devil shadows him forth, as ubiquitous or high, then it would be harder to find what is the seed-fact of idolatry. But if we are to worship God as holy, and His whole right over our worship springs from His character as excellent, then we start into idolatry when we drop even for an instant the holiness from God.

Then another eventuality subsists. Dropping holiness, we drop reason. This is clearly revealed. We turn the truth of God into a lie, and God gives us up to birds and four-footed beasts and creeping things.

Observe these two facts:—(1) Idolatry begins ethically; that is, its germ is set when we turn away the least from holiness. An impenitent man is a commenced idolater. But idolatry is very thoroughly matured when we bereave GOD of holiness. (2) And when this occurs the second stage immediately follows. We are given up to old wives' fables. And reason, which might seem but fleshly, is stamped at once with the error we have made.

CHAPTER III.

Fetich.*

Fetich bears this last imprint. It is not only wrong cult, but beastly. If I not only de-moralize God, but turn him into a pig, I grow irrational. All idolism is more or less irrational, but it grows more so as it seats its curse. There springs a miracle of what is stupid. The classic Greeks shall worship Priaps and the Eumenides. Great Pharaoh shall bless an ape. And there shall supervene a crazy streak upon the intellect of the most cultivated of men

CHAPTER IV.

THE TWO ATTRIBUTES OF FETICH THE TWO ATTRIBUTES OF THE ABOVE DESCRIBED TEN DOCTRINAL PROPOSITIONS.

The two attributes of Fetich are the turning aside from a holy God, and the being cursed for

---

* Fetich would seem a wrong word; for a common dictionary would call it a material idol. But we prefer it for two particulars, (1) first it is the *system* as well as an idol, for it is used for the charm-power, as well as for a gree-gree or a shark's tooth or any one fetich. It is a sailor's word, at best, put upon the natives for its sound, like their *fetisho* or Portuguese *feitico*, which means charm-system as well as charm (see Le Brosse). Moreover, (2) it has vast resemblances with us, for it is the Fetich of a people who believe, as we do, in a Nzambi or Great Divinity. They catch up their fetiches in random ways from almost anything; and though doubtless, this origin, as of the bronzes in India, does not forbid with the low people a more direct adoration, yet this mixed state fairly illustrates the mixed state and all the more senseless conditions of cult-thought among ourselves.

that by the most extreme irrational mistake of Him. The two attributes of those statements were the emptying of holiness out of God, and then, just as Paul threatens, the paralysis (most symptomatic in its look) of everything rational. Can the resemblance be the birth of chance? How is that possible? Can our Church be impervious to such mistake? Why should we think so? Is it too intelligent? Look at the calf and the crocodile of the high period of Egyptian civilization. Is it too pious? Look at the primitive age : and yet the mystery of iniquity did already work. Are we too sudden in this brutal lapse? because *Nemo repente turpissimus?* But look at the long years of younger teaching. Dr. Hodge does but mature the work. We firmly believe that the origin of East African superstition and the origin of these dogmas of the Church are identical and one, and that it behooves us by everything that is of the truth to show that these things are not Calvinism; at least, that if they began in the Calvinistic books they are not Paulinian; nay, that they are not of Calvin in the higher sense of the Calvinian thought; that they are not of Dr. Hodge in his better teaching; that they are not of the Holy Ghost ; that they are ripening and crusting over and endangering our orthodox belief; that they are tumors, and not of the organized flesh ; that they can be cut out, and the creed will live ; and that it cannot be a day too soon when we come back to this,—that God and man are in each other's image (Gen. i. 26); that holiness is an intelligent trait; that holiness is the secret of our worship; that holiness must be the centre of our creed; and that when ten dogmas take the high places in our Temple, and are exempt of holiness, it is a sign that the Church is tainted with Fetich, and that she is to get her curse of it in the direction of her *gangas* or doctrinal chiefs.

CHAPTER V.

HOW BEST TO MAKE THIS APPEAR.

WE are so desperately weak when we utter such things, that it becomes a serious question how we can manage best; for though we manage our very best, we are sure to outrage most men's most settled convictions, and so have all the sting of managing badly.

Most men will laugh at us. Many more will turn away, and not read a sentence after the first. All will have some tincture of disgust. To take a gree-gree of the Congoese, and say,—That is a type of long cherished doctrines; to go into our last Theology, and take from the very cream of its results doctrines of their very nature central, and hold them up as a mistake, and compare them with the very grossest superstition, seems too mad to be mere impudent affront, and to belong rather to the region of queerness and infatuation.

What would be the shrewdest course for a mind awake to results, if he wished to avoid unnecessary denunciation? Not to stop, of course. That would be wicked. God's making everything for display is a grosser thing for a polished philosopher than the fish-skin of the Guineaman, and there are the common absences in both,—first of holiness, and then of rationalness,—betokening a common print of kindred superstition. How ought we to act? Pause and consider and be sure we are right, and delay for years the scandal of such hazardous opinion. And, after that? Why—broach it in the most careful way.

And in trying to find what that is, we are led by instinct to consider, first, *Fetich in Practice*. There certainly the Church is Idolatrous. And as the Church does not think herself idolatrous, and rarely speaks of it, and in all this wide province of sin rarely makes any confession, perhaps that may begin the suspicion. Some may agree that there may be a great deal of idolatry, for the cause and by the very reason that we are unwarned of any.

At least, then, we are idolatrous practically.

And having established that, past all possibility of mistake, it will make us stronger about *Fetich in Doctrine*. We will afterward go further still, and find we have *Fetich in Order*. And these, therefore, will head our remaining books :— FETICH IN PRACTICE, which we will have no fear not to make thoroughly confessed ; FETICH IN DOCTRINE, which will involve the Ten Doctrines which we are more thoroughly to discuss ; and after that, FETICH IN ORDER ; showing that the same bony finger of superstition is pushed,—and for the same cause,—to shrivel and spirit-away the Church.

BOOK IV.

FETICH IN PRACTICE.

CHAPTER I.

THE BIBLE A FETICH.

God may become a Fetich, but only by ceasing to be a real God, and by becoming a false Divinity. The Bible may become a Fetich, and remain a real Bible: of course we have no other. A real Bible is a wonderful book both in the province of reason and holiness. In the province of reason it has a geography of its own. Where this book sells itself, print and paper and calf-skin though it be, there is the world's nobility. Take a pencil, and go to a common globe, and draw the lines within which the Bible is read, and those portions of our planet can sell and buy and govern and out-think and out-general all the rest. Reason, even where it derides the book, yet seems to nestle in lands where it is kept and printed.

Holiness has a still more wonderful geography. It has a Land where all are holy, and an Age which is to last forever; and in all the wonders of that Land, whether of mind or matter, whether of intellect or conscience, there is no inhabitant that has not been made such by the Bible. Even Seth and Enoch had some chapters of it. And here is a book, manufactured of pen or type, that the millions of the saints know something of in the work of their salvation.

Now, it is the religion of Protestants. But how easy to let reason and holiness go out of it, and to use it afterward for a wretched fetichism.

The mother gives it to her son; and in a war or on a distant journey hides it in his trunk, or binds him to read it daily.

He does so.

I do not mean that he ought not to do it, or that the mother ought not to do all she does, and to secure by all means this daily use of it. She ought, beyond all question. But what supervenes? Why, in a thousand instances, a mere fetich. The boy reads, and it is his idolism, like bowing to an image. He does not feel, and he does not think, and he does not remember. Ask him, and he does not know what he read. He stands up by the gas at night, and reads a chapter. It is his cult. It is the tribute he pays each day to his religion. And he throws himself upon his bed, empty of anything that has been meant except a growing need of some just such idolatrous superstition.

And so of Prayer :—

CHAPTER II.

PRAYER A FETICH.

PRAYER is, perhaps, half of piety: I mean by that,—Prayer is, perhaps, half the means that a man has to build up his religion.

Prayer is, perhaps, half our usefulness. The twelve, when they appointed deacons, said,—" We will give ourselves to prayer, and to the ministry of the word" (Acts vi. 4). They place prayer first. Prayer, perhaps, is half of any man's possible usefulness. No wonder, therefore, that men value prayer.

The same boy that keeps a Bible by his gas-light, dashes down afterward, and says his prayers. But ask him three minutes afterward, when he is in bed,—My son, what did you pray for? and he will pass his hand over his eyes,—and he will be puzzled to say. Recollect, this is his prayer. His mother has been bent that he should make it. Recollect, it is talking to the Almighty. If there is anything more solemn on the earth I scarcely know it. And it is because it is so solemn, he prays; just as because it is so solemn, the Congo-man keeps hard tied-on his shark-tooth Deity. But notice the elements that are absent. (1) There is no holiness in it, for he is a dissipated boy; and (2) there is no reason in it, for he merely mumbles it. It is his fetich. So here, in the very heart of what is orthodox, there is a distinct and confessed idolatry.

CHAPTER III.

SERIOUSNESS A FETICH.

MEN, when they get religion, become serious. Men notice that, and it becomes an index of religion. They make a list of who are serious.

Man is a proselyting enginery, and if men stay serious we are fain to think they are religious.

We will not paint too tediously. What bystanders encourage the man himself may eagerly consider.

There are certain gestures of the pious,—a certain motion of the eyes and a certain raising of the hands and a certain settling of the face, which men learn, and which in their appearances in church become the *indicia* of piety. What else can the Church generally judge by? These symbols react upon themselves. They deepen and become more express; they settle and become inwrought; they characterize, and give identity to, their possessor. Like all outward signs, they grow too valuable, and are mistaken for something else; and, unless open rascality unmask the fraud, the man may wear, like a fish-skin about his neck, his gravity as the whole of his salvation.

CHAPTER IV.

PROFESSION A FETICH.

The Bible had peculiar idioms. In Eastern lands men bargained by their speech. They used outward symbols. Their merchants were travellers,* and, therefore, in nomadic ways made bargains with their mouth and less with the solemnity of pen and record. Hence human conduct came to be depicted by the forms of utterance. Paul says,—" If thou

* Travellers, therefore, came to mean merchants (Gen. xxiii. 16).

shalt confess with thy mouth the Lord Jesus" (Rom. x. 9). Christ says,—"Whosoever speaketh against the Holy Ghost" (Matt. xii. 32). Solomon declares, that "death and life are in the power of the tongue" (Prov. xviii. 21); and all through the Proverbs a man is said to have "joy by the answer of his mouth" (xv. 23).

We have drifted into other modes of putting the matter in Western languages; but, alas for us! we keep the snare which such an exaggeration of the power of speech is fitted to convey.

Profession is a terrible snare to us. It puts us down among the saints. It causes us to be treated as though forgiven. It moves towards us ghostly ceremonies. It embalms us in the cerements of the blessed.

Who ever repents after a profession?

The Church may shake our hold if we disgrace her. But if we are moral men, like Jews and Papists we have an established place, and it is hard after that to reach us as impenitent.

And yet how lightly we professed!

When we remember how sanguinely men sometimes join the Church, and how they promise themselves afterward to make up what they feel they need, and how profession grows to them and becomes a badge of grace, we look at this as perhaps more than anything else showing how the orthodox may have a fetich, and how holiness may be thrust aside, and a mere act rank it in the number of our evidences.

CHAPTER V.

ALMSGIVING A FETICH.

ALMSGIVING has either of two motives,—either, first, benevolence, or second, a desire of gain. Almsgiving, as benevolence, is itself of the very nature of holiness. Almsgiving for gain, is, of course, nothing of the kind. But see how this descends into the most insidious details. Men may do alms with no notion of display; they may have gotten-by that; nay, for no worldly recompense. It is astonishing how fine the influence may have become. They may seem the most serious and the most austere believers. They may seem entirely devoted as by their inward sacrament. They may have heard of recompense, in heaven, and may not give alms even for that high-set and far more distant retribution. And yet they may not have a spark of piety. They may give alms as of the nature of the proof that they are pious. They may give alms as part of the machinery by which they hope to be forgiven. They have been warned that they must not do it for recompense and, therefore, they do not do it for recompense, even as brightening their crown and increasing their bliss among the sanctified. But they do it to be sanctified. It is astonishing how near we can come without our real motive being holiness. They do not do it for benevolence, and they do not do it for their love of right, and they do not do it for a desire of right in the sense of being made holy, but they do it as required of piety. They do it as they go to church, and as they wear a long face, and as they witness a fair profession,— as what they are taught as the A. B. C. of works, and as what they are willing for as the condition of salvation.

There is nothing more terrible than how the ship can be laid near the port, and how indistinguishable the phases of possible deception.

The test of piety is holiness. All else is fetich. The closer character can be brought to this (I mean when it is articulately another thing), the more dangerous the mistake, and the more serviceable, in the symptoms that it offers, in the diagnosis that is yet before us.

CHAPTER VI.

PREACHING A FETICH.

IT was, perhaps, thirty thousand sermons that I saw marked the other day as preached every Sunday in the Islands of Great Britain. This makes fifteen hundred thousand annually. It was said that as many were preached in our own land.

Preaching is a wonderful institution.

And yet Paul never preached a sermon.

We value the chance *in distinctu* of saying a word on this great and most venerable, and yet fearfully misquoted, and dangerously overestimated, usage of our religion.

Paul praises *the Church*. He says it is "the pillar and ground of the truth" (1 Tim. iii. 15). He means by that that it treasures it,—that it is its Great Librarian. He ennobles the *parents'* work, and makes it vital. "Children, obey your parents in the Lord for that is right. Honor your father and mother which is the first commandment IN promise." When he gets into the arena of the word he takes care to say that it is to be promulged in every possible way. That he took a text, and preached in our style of discourse, we have no knowledge; but that he taught in all possible forms; writing letters; making public harangues; speaking late at night, as when the young man fell from the window; disputing daily in the school of one Tyrannus; talking to soldiers tied fast to him; standing before Felix and Agrippa; and seeking interviews with families, and with them of Cæsar's house,—we know; and when he talks on homiletics, he talks in like way. He tells Timothy to be "instant in season, out of season" (2 Timothy iv. 2). He seems designed in Providence to break up all chance of homiletic superstition. He tells Timothy to "reprove, rebuke, exhort with all long-suffering," and, what is very express, "all teaching." It would seem impossible that we are to glean from this one recipe of speech.

And yet what has been the Church's inference?

Most solemnly this,—That the sermons on the Sunday are the special instrument of salvation.

We hear a minister say that he has converted two thousand souls.

There have been pious parents, and countless

ministrations of the church, all of which Paul praises as the very pillar of the faith, and yet some semi-centennial will give a positive list and say, that two thousand three hundred and sixty-five souls have been some old man's spiritual children. He may be a miserable preacher. He may be covertly known as inefficient in his gifts. He may be secretly felt to be an incubus on the house of prayer. And yet, simply copying from the roster of the clerk, he sweeps all in as though they were saved by preaching.

And now the warrant!

Paul does say,—"It pleased God by preaching to save them that believe" (1 Cor. i. 21). But mark the family likeness. Christ says,—"Except ye eat the flesh of the Son of Man" (Jo. vi. 53). We are told,—"Baptism doth now save us" (1 Pet. iii. 21). How do we guard against those idolatries? It is *easier* in the instance of preaching, because κηρύσσω, the very word itself, had no technical phase in the days of the Apostle. It meant heralding in all possible sense. When he wrote, he preached. When he rebuked Peter, he preached. When he drew out formal systems, he was a κῆρυξ in the sense of our Redeemer. And, therefore, in the original charge,—"Go ye into all the world—and preach" (Mark xvi. 15), there immediately stands the gloss, of διδασκάλειν or making disciples (Matt. xxviii. 19).

Better say,—Parents have done the work. Ministers have turned the hearts of parents to children, and in ten thousand ways of Sunday Schools and admonitory teaching the Church has elevated her saints. This sounds like rationalness. The old man's fantasy sounds like fetich. We simply challenge any right to it. If Paul himself never preached; and I press attention to that: if he leaned all upon the word, and asked,—" How can they hear without a preacher?" but made himself specially versatile and maniform in his talk, and threw himself into every attitude for discoursing on the gospel, how crusted our pulpit mode, and superstitious the claims that are to be built on its Sunday ministrations!

(1.) They breed contempt. Men see that they trace influences to the wrong efficiency.

(2.) They breed neglect. Others turn from their own private duties to the ministry that is to work so much.

(3.) They breed delay. Men cease all lower methods in this zeal for the sacred desk.

It has all the symptoms of the gree-gree and the Ebo. For though it is glorious—this modern habit of sermons in the church, yet all our use of them beyond their value as a most rational arrangement of the Word, is just as much a superstitious wickedness as the shark's tooth and the eel's skin of the Nzambi worship.

And this is most fatal when transported among the heathen.

We were present years ago in a Baptist Court where it was decided that there should be no more schools among the heathen. Large schools in

India had long existed; and the missionaries who had originally founded them were present, and had come for their defence. They had presented in the most touching ways the evidences of ·God's favor to their work. I was amazed when any one rose on the other side. The argument was that preaching was the means of grace. As though it were some charmed touch, men argued as though we had only to leave it to the Almighty. Though it might be in stammering speech; though it might be irrational to dream of its success; though it might be hardly understood, and, from difference of vocal idiom, wretchedly out of place, yet men talked as though it should be followed up at fairs and bazaars, and as though we should scatter it as it was, whether men might hear or whether they might forbear.

I was still further amazed when these views were those of the Court. And when grave missionaries of Christ wept over the vote, and finally declared that they could not sacrifice their usefulness, and gave notice of their withdrawal to glean what they could and to labor under the care of certain English friends, we received an impression that we have never lost, of Fetich in the orthodox Church.

We heard the like argument but the other week, that it was wrong to have schools among the heathen. It was by a venerable Presbyterian. "Schools may look well to human reason, but God has chosen preaching. We have no right to go back of God. Peter obeyed God, and preached, and converted three thousand men on the day of Pentecost (see Hodge's Theol., ii. p. 345). What if he had been an advocate of schools?" And so benumbing are the influences of these sorts of long growing superstition, that men are petrified, as they are said to be under the mesmeric touch. Who said that Peter converted three thousand men on the day of Pentecost? We are expressly told the opposite. Who said that God had chosen preaching? There was no preaching in any sense that was exclusive for hundreds of years. Who said that we were to scatter preaching, trusting to a miraculous power? On the contrary, God had ordered for it by the most rational means. He had made the whole world a school. He had ripened it with an Augustan wealth. He sent all over it the Jews. He had filled it with Jewish synagogues. What was RATIONAL could hardly more have been attended to in preparation for the coming of Christ. And when Peter preached he preached not at cross-roads and to heathenish pilgrimages, but at synagogues. The Lord had graciously prepared for him by religious schools. And on the day of Pentecost he was not preaching to Pagans but to saints. The largest number were perhaps intelligent believers. It is not certain that he converted one of them. For we are told that they were Parthians, Medes, etc., *devout men* out of every nation under heaven (Acts ii. 5, 9). What a noble preparation for any enterprise! And now, when we go among the heathen, and neglect what our Lord says,—" Go not from house to house ;" when we fail to see in this what its quaint image meant, that we are to begin like a fire on a hearth and smoulder gradually outward; that we are to have our rational attempts; that we are not to go among brutes, and throw religion at them as Peter did not do even on the day of Pentecost, trusting that God will bless His own word; that we are to remember that Paul went straight to synagogues, and had no such difficulties as we have on heathen shores,—we are unready (even without Paul's miracles to help) to use Paul's humbler means for the building of the gospel.

Schools, men suspect.

Forty years ago if we had relied on schools, where should we be now? If leaving Hawaii, where there are few to teach, and where the race decays, we had gone to broad continents, and laid deep schemes like the children of this world; if we had erected Paul into our beacon, and coveted Paul's "synagogues," and created something like Paul's synagogue-schools, God Himself would be more in league with us; for He follows just these schemes of influence. He begins low down, even with the young. He builds slowly out like the forest from the seed. He works slowly forward, beginning with the very least; and hence the huge results of these compacted and unwasted increments.

We have idealized a figment; for we have worshipped preaching, and preaching, when we come to see, is really no one thing at all.

And, furthermore, the Church's prayers and her ministers' earnest labor having been promised a reward, the very nature of the reward is indicative of the Church's blunder. God rewards by pushing the Church's work out of the way. By colonies in Africa and from among the Chinese, by changes in Japan and among the Papal States, by civilization, as in Brazil, and by steam-travel and telegraph alliance all over the earth, God has listened to the prayers of His people, and hinted that these secular results should be imitated more than they have been in the conduct of the Sanctuary.

Brahm is intolerant. He allows no usurpation of his place. Preaching is the appointed means. And if this is exegetically a mistake, and κηρῖσσειν takes in all plans even down to godliness in our personal behavior; then preaching is a fetich, and glorious means for the extension of the truth have been paralytically held back for these Sunday ministrations.

CHAPTER VII.

FAITH A FETICH.

WE have had some executions lately which rouse a suspicion, which we wish to think of, that faith itself may be idolatrous.

Conversion may be at any hour, and of the most desperate of the wicked.

This we agree to.

Conversion of whomsoever, and out of whatsoever of which we can form any judgment, and wherever as concerns time and place, is a thing that we are to consider possible; only we hold that this amazing mercifulness makes more critical the question of the reality.

When a man murders hellishly, and goes to jail with the most profound indifference! when he trusts his lawyer and a blunted court, and lies fearfully as a method of deliverance! when he feigns insanity, and curses Christ, and continues this to the very edge of execution! what are we to think of a faith that springs up out of the very soil of horrid blasphemy in time to save him? What are we to think of whole rows of such events?

Is there not a flaw somewhere in our theory of faith?

The Israelite was made whole by simply looking at the Serpent; but then the serpents had a fiery sting, and he had to *feel* that, before he would consent to be delivered. Do these men feel the sting of sin before they tell us on the gallows of their certainty to be delivered?

What strange spectacles those have been—men convicted of the meanest crime telling everybody that *they cordially forgave them!* There is a likeness in each case. The feeling does not seem to be a withering conviction. They are not shrinking from the gaze and shuddering under a sensitiveness to their horrid iniquity, but they are thinking of themselves. Their hope seems to be a willingness to die, and their uppermost thought what they may look for when in heaven.

May there not be a horrid idolatry in this?

Ministers lend themselves to such scenes, and are the very *point d'appui* of this ghastly comfort.

The shudder of the mob, and instinctive gainsaying at such spectacles of confidence; the incredulous shaking of the head, and smile of bated disrespect;—what does all that mean? Wickedness perhaps; but then, to us, a ground for legitimate suspicion. Can it be that these weekly gibbets reveal a faith that was intended in the preaching of the gospel?

We believe that faith may be a convenient fetich. If we would characterize that we would say, that it is neither (1) holy nor (2) rational. Regeneration is a moral work. We cannot believe till we are regenerated. If we believe, we are penitent. We cannot believe that these are accidental facts, but that the nature of faith is moral, like being born again.

What moral faith has been bred upon the gallows? The noblest, if it be the work of God; the humblest, if He has decreed salvation; the most real as within the possibilities in the case; but what are the probabilities? If it be known that there is a most cunning counterfeit, and, moreover, the ingredients are all categorically given;—first, fear, melting a man like wax on the edge of the fire; secondarily, knowledge, telling a man all he needs in respect to a complete salvation; thirdly, hope, and fourthly, despair, giving him no other hope, and shutting him up to mercy in the gospel plan; what part is wanting of a high and earnest faith that could at all distinguish it? None, certainly, except the moral part.

And of this we can detect the absence, when we watch him narrowly. He is thinking of what he is to *receive;* not of his wickedness. Pardon and new born peace and fresh-bought right to happiness rise upon his lips. And there are priests of a hypocritical religion (ignorantly such) who help him. These men want to know if he is willing to die; and if he die triumphantly, as hundreds do, the scene of that triumph seems to carry all away from the question whether there has been any shame for sin or tender brokenness of heart for all he has committed.

There must be a better faith.

Religion is a total renovation (the Bible is express); a new birth; a rising from the dead; a second time to be created. It is the change of all imaginable changes,—out of darkness into light. Faith must come out of it. To take a man's head upon our lap when he has been dashed out of a broken rail-carriage, and whisper that he has nothing but to believe, is all true undoubtedly, but Oh be careful how we emphasize that word, nothing. Everything may be transacted; that I know; everything forgiven: I do not dare to say that everything may not be done, as the blood pours out upon the grass. But I do say,—Satan revels at such death-doors, kindling the very readiest deceits, and encouraging the vile who are there to witness it.

Religion is moral. Faith is an act of it. Pardoned, I am converted. Conversion goes down to the very bottom, and is a moral revolution of heart. Faith, therefore, is a moral vision. And if the soul, come to the brink, has but two minutes till it pass into the other world, I must make that moral fact appear. Faith is a new vision. And all I tell the sinner, as his soul is passing away, is to look to Christ. Christ can give him that moral apprehension. And yet I must tell him, Christ *must* give it, —*will*, but *must*,—*will* give it if he cry earnestly for succor, but *must*, or the soul is lost. A repentance which is an act of faith, and a faith which is an act of penitence, must both be given, or the soul cannot enter into the presence of the Master.

The want of this is peopling the church with hypocrites.

And if any man says,—There is no time for this. This poor soul is bleeding out upon the turf; behold, now, the very nature of the fetich! how it tempts men to put off the day of grace ; and when the danger comes, narrows in the faith to a mere glimmering of the old intelligence.

Let all men cry for pity, and when they cry, let it be known that Christ will hear them, but while that cry *haeret in cortice*, let it be known that it is a mere burst of terror. God will hear a mere out-cry of terror. But let it be known, when it is no more than this, it may be precious, but it is not faith. And when it is not thought of till the last, it may be but the selfish brokenness of a soul in its horror when it comes to perish.

4*

CHAPTER VIII.

REPENTANCE A FETICH.

REPENTANCE, therefore, may be also a fetich.

It is astonishing what a counterfeit may spring from spiritual terror. Fear is not a part of religion, —in fact, perfect love casteth out fear; but fear is a "sorrow for sin," and in a very thorough and honest way there is scarce any greater "change of mind."

The victim on the scaffold feels like an altered man; and when the ministers are speaking and press hopefully the methods of salvation, his is a new life, which may break forth into the most joyful shouts, and into the most tumultuous agitation, as he waits to be launched away.

How can I say to such a man anything wiser, even though it be in the article of death, than that repentance is moral?

And so of all the things of which we have been speaking, the Pulpit, and Faith, and Repentance, if I let them drift and separate themselves, one desolates the other. If I make the pulpit mystic, men send their sons to it as they do to a doctor or to school. We remit moral nurture. Here, in the spired synagogue, is the charm that brings deliverance: life at this rate is always at a crisis. And caring less for morals, and more for revivals and great rare occasions in the sanctuary, we adjourn personal watch, and treat our virtues as though they were mere insubmission to the gospel.

Then faith crowns the building. Treat that outside of holiness, and the mischief becomes complete. Religion becomes an orgy like the phantasms of the Greeks,—like some bottom land producing a ranker jungle. Would you see the symptoms? look around: believers in the Church, rascals in the market-place; men's minds made selfish even by the gospel as a scaffolding for heaven, like Babel, to lift them above the deluge; Christ a convenient rock; faith a mere carnal scheme; repentance a mere trading debt; and a thorough radical reform no part intrinsically of the doctrine of salvation.*

CHAPTER IX.

THE RATIONALE OF FETICH.

BEFORE proceeding next to FETICH IN DOCTRINE we wish to interpolate another step in our analysis, viz., the Rationale of Fetich itself. We have seen (p. 56) how the man of medicine must have a double search, first, after the symptoms, and second, after the disease that gives them nourishment.

* Notice how near akin Doctrinalism is to Ritualism in this respect of immoral tendency :—" It may seem difficult to reconcile gross deviations from morals with such devotion to the cause of religion. But the religion taught in that day was one of form and elaborate ceremony. In the punctilious attention to discipline, the spirit of Christianity was permitted to evaporate. The mind, occupied with forms, thinks little of substance. In a worship that is addressed too exclusively to the senses it is often the case that morality becomes divorced from religion, and the measure of righteousness is determined by the creed rather than by the conduct." *Prescott's Conquest of Mexico*, vol. iii. p. 362 (Ed. of 1847).

He has a third search, viz., after the occasion that begets the disease. First, the symptoms; second, the disease, and third, the occasion, or why the cold or the fever throws itself upon these particular viscera: these are our stages.

One symptom is irrationalness. The disease is idolatry. Now God threatens idolatry with irrationalness, that is with judicial blindness and insanity of life. Still, judgments have their causes, and I am free to approach the third stage of the inquiry, and ask why, when holiness glides out, reason follows, or why in a most refined age man may be most astute in general, and most senile in the particulars of his worship.

Is there not an analogy in outward things?

If I spare my legs, I improve my arms. If I lose my eyes, my touch gets an exaggerated gift. If I am smothered out in every other sense, my ear becomes ten times more vigilant.

Then so in the province of religion.

If I throw holiness out, I must exaggerate other things, or I will have nothing plausibly to worship.

All sinfulness is at heart an idolism. For look at the stages:—First, practical unholiness, which is the trait of everybody;—second, doctrinal unholiness, or holiness gliding out of our creed; third, an emphasis upon something that shall supply its place; and fourth, irrationalness; as this substitutionary trait is of need preposterous and brutal.

I need the idol.

The Rationale of Fetich.

I cannot worship nothing. If I take out holiness, I must put in something different. If the Greek abandon theism, for God's sake he must have some divinity. The soul yearns somewhere. The bronzes of the Hindoo! They were symbols first. They became idols afterward. If I take out what is osseous I must fill up with degenerated cartilage.

Hence the extreme *irrationalness* of idols. Sin was the *proton-pseudos* of the degeneracy. Admitting sin, and so expelling holiness and driving that holiness, therefore, away from adoration, and worship had to fill itself up with what it could; and, therefore, cats and crocodiles were but the extremer refuges after a forgotten holiness.

This sheds wonderful light on our theology. God, not one thing,—then He must be something else; not holy—must be everything beside more plentifully. This is what begets deformity. Loveliness being taken out, a mere Goddy-god must be put in; a mere shell-like Deity. There must be mere arbitrariness, the crust of a hearty worship. And while we obscure the wound by speaking of Him as an innate idea, yet there is a grim look on what is left, like the glassy eye that we try to think of as the very effigies of Nature.

God bereft of holiness is bereft of rationalness. That He be bereft of rationalness is a special judgment upon the Church. And yet, nevertheless, though this be the testimony of Scripture, it is a special judgment by force of reasons in the case, and by stress of facts that are resident in the necessity of nature. This will help us mightily in the discussion that is to be the next to follow, and where we think this irrational trait is wonderfully exhibited.

BOOK V.

FETICH IN DOCTRINE.

IN bringing so hard a charge, and approaching the crisis of our book, we had better proceed cautiously, and we will observe this usual method:—First, quote Dr. Hodge, that he may give his own doctrine; second, quote Dr. Hodge, as he promptly contradicts it! This is a strange peculiarity. It is a Providence and a fact, but it is hardly an accident, for it is too punctual; or a mannerism, for a mannerism is a thing of style, and style could not so affect thought. The contradiction is often more extreme than those not guilty of the belief would think of or concede. Nor is it a necessity; for though truth does dispose of its mistakes, and a great system raise the walls of its impracticablenesses to bar out inadvertences in itself, yet that is more gradual. Here the suicide is at once; like a bee-bird close upon the bee. Hardly has Dr. Hodge let loose one of these ten propositions from his pen, than he remorselessly, and I may say uselessly, unsays it. Not *in eundo*, or in the necessary tread of his polemic: that would follow indeed; for there is enough of glorious gospel in the theology of Dr. Hodge to make it wonderful, as we see lucidly through it, that, as in Maelzel's Chess-

Player, so much that is merely human can lie concealed. But what we mean is fatalistic, a hap like a bird devouring her eggs. Not chance, for we have learned to expect it, like the rumble of the earth followed by the tidal from the sea. This, therefore, will be our second stage,—to wait for the tide that shall sweep each first proposition.

Our third shall be to consider Dr. Hodge's texts; our fourth, to offer our own; our fifth, to consider the argument from reason; and our sixth, to finish the discussion in each case by tracing the proposition in hand to the soul's idolatry.

Let us, now, take the ten up in the order in which they were propounded.

CHAPTER I.

A GOD ALL FOR HIMSELF.

§ 1. *Dr. Hodge's Statement of the Doctrine.*

Dr. Hodge says (Theol., vol. i. p. 436),—" If we make the creature, and not God, the end of all things, our theology and religion will be in like manner perverted." He says (i. p. 535),—"all things are said to be not only of God and through Him, but for Him." He says (p. 566),—" It follows from the nature of an infinite Being that the ground (i. e., both the motive and the end) of the creation must be in Himself. As all things are from Him and through Him, so also they are for Him." He says in the Commentary to the Romans (iii. 25, p. 129), —" God is the ultimate end of all His own acts:"

(Theol., vol. ii. p. 321), "As He is the Alpha, so also is He the Omega; the beginning and the end:" (ii. p. 339), "This appears first from the clear revelation which the Scriptures make of God as infinitely exalted above all His creatures, and as the final end as well as the source of all things."

§ 2. *Dr. Hodge's Contradiction of his own Doctrine.*

Intertwined with these sentences and occurring very much oftener than they do, are others which teach the doctrine which is the fifth to be considered. By that, God is represented as making it His chief end to show His Glory. We wander from one to the other without being taught to consider any difference between God's self being His chief end and the display of Himself to the eye of the beholder.

The haziness that comes over our sight is not in the least relieved.

God's chief end is of necessity positive. It is clean-shaved. We have a right to think of it as unique. To tell us, first, that it is Himself, and, second, that it is His own display, is certainly to divide things at once; to make things definite by two ideas; and to make those ideas specially unmeet by choosing one to be the Great God Himself, and the other the temporal end of showing Him to the creature.

If I say,—A man makes everything for himself, it would strangely mix things to add,—He makes everything to show himself. The objects are not equipollent. I may explain by saying,—That man loves showing himself more than anything in the world beside. But still, this latter object is a means. The man's great end is himself. His delight is in his own display. His great object, therefore, is not the exhibition, but himself, the exhibition being but the means whereby he furthers the great end of his being.

But there is a worse divaricating. Hardly is either doctrine broached before Dr. Hodge is eager to defend it. Why? If there be a highest motive can I vindicate it by a higher? Standing forth as the last end must it not be topmost and in itself? And yet Dr. Hodge goes on to topple it and to seize its place by showing that it is RIGHT that it should be the chief end of God.

If God's chief end is absolutely Himself, that is the term of all possible motive that we can suggest. To palter afterward is to recant. To say that it is His own display is mixing matters enough: but to grow timid and to bring in the motive that it is RIGHT (i. p. 567 and ii. p. 339), is to confess judgment in the very plea, because it is to bring in a higher object than self, and to make rightness *ipso facto* the regent over all.

We are to return to this subject under the fifth head, and, therefore, we will not now quote at length lest we recapitulate too much, but one doctrine, in our belief, being all that is necessary to take the place of both those that are propounded by Dr. Hodge.

§ 3. *Texts to Sustain the Error.*

The doctrine that God has made all things for Himself seems to be announced by Solomon, *ipsissimis verbis*, in the Book of Proverbs (xvi. 4), "The Lord hath made all things for Himself." This doubtless is the text from which just this form of statement has been taken. But then not only are we too prone to make superlatives in Holy Writ of what is far more moderately stated, and to speak of the *chief* end and of the *highest* motive when only one end of God is legitimately to be inferred; but Rosenmüller shows that this whole expression is entirely ambiguous. The pronoun may be either *Himself* or *itself*. If it be "*Himself,*" it refers to God; but if it be "*itself,*" it refers to "*everything.*" "The Lord hath made everything for itself." Maurer believes that it refers to "*everything.*" But Ewald comes in still further to divert the sense. He shows that there is an *article* in what has been supposed to be a compound preposition. He shows that that *article* draws back a substantive which otherwise would be used as of the compound preposition. He shows that that substantive is eminently in place. Solomon had been using it. The Proverbs just before contained it. It is from the verb to *answer*. It means a *decree*, or that fearful *answer* that a man makes to some life question or request. "A man has joy by the decree of his tongue," says the last chapter (xv. 23). "The preparations of the heart in man, and the answer (or decree) of the tongue is from Jehovah" (xvi. 1). This alters the prooftext entirely. "Jehovah has made everything for His decree." That is, before time He gave an *answer* to the whole question of His eternal Providence. That answer is His decree. That decree is complete for His whole eternity. "I know that whatever God doeth is for the universe" (Ec. iii. 14), says this same Solomon: by which I do not understand, as Dr. Hodge might, that this is His *chief* end because it is so discoursed of, but that it is *an* end. God doeth everything for the universe: which is exactly tantamount to saying,—"[He] hath made everything for His decree:" the meaning being that He had a whole plan at the very beginning, and that "even the wicked for the day of evil" were woven as a vital part of it.*

So Paul's text (Rom. xi. 36),—"For of Him and through Him and to Him are all things," which has been perpetually quoted. This wilts at the first touch. "*Of* Him and *through* Him," of course, are quite impertinent. "*To* Him," must bear the exclusive stress. "*To* Him" is all to be interpreted by the uses of the preposition εἰς. Now look at that word εἰς. "They were all baptized *into* Moses," i. e., *in reference to* Moses (1 Cor. x. 2). Such is the commonest of the tropical significations. How idle to build heavily upon such a particle! "Of Him and through Him and to Him are all things." That is, they are "of Him and through Him" as Creator, and "to Him" as Guide and Judge. It is cruel to

* See the Author's Commentary on Proverbs, *in loco*.

tempt men upon so thin a plank, and then leave the intellectual and the wise to suppose that such are the Scriptures that we are to depend upon for the most serious gospel.

For look further. Dr. Hodge takes an expression of Christ, " I am Alpha and Omega " (Rev. i. 8). He lays great stress upon it. Think what it is to prove !—So thorough a thought as God's uttermost end ! And yet look at the language. He is the Omega in a thousand ways. The analogy of Scripture, and all the broader principles of truth and right, seem to be as nothing to Dr. Hodge before one of these flying texts ; and when we come to examine it, John himself takes it quite out of Dr. Hodge's hands. It might mean one thing. It might mean another. It might mean anything answered to by the emblematic imagery of what is final or at the end. And, therefore, Dr. Hodge *assumes* that it means *end* in the sense of *object*, and that it means, too, the *chiefest* end or that which is absolutely *highest* in the conception of the Most High ; and yet we bend and listen, and the Apostle has already begun to explain his own sense, and Dr. Hodge is already away with his assumed and divergent signification.

We do appeal most earnestly here. Ought not high theology to set the pattern of grave and most careful vindications ? And yet here is an instance where John has gone on to interpret :—" I am Alpha and Omega, WHICH IS AND WHICH WAS AND WHICH IS TO COME ; " that is, whose being is eternal ;

while a theologian has already caught it up; turned it in another sense; and taught from it God's purposed end, and that the chiefest one, in the work of His creation.

§ 4. *Texts to Refute the Error.*

Elihu says,—"If He set His heart upon Himself, if He gather upon Himself His spirit and His breath, all flesh would perish together, and man would return to the dust" (Job xxxiv. 14, 15). It is true this is not the usual translation; but we will risk it; the *Margin*, even, suggests it; the Hebrew ordains it; the sense obviously demands it. It is the doctrine of the whole Bible. God is everywhere sufficient in Himself. Is it any gain to Him if we are righteous? (Job xxii. 3.) "I know," says Solomon, "that all that God doeth is for the universe" (Ec. iii. 14). And then he says, more flatly,—"God doeth (or acteth: "*it*" is in *italics*) that men may fear before Him." To this belong all those texts where character is spoken of.

"Justice and judgment are the habitation of [His] throne" (Ps. lxxxix. 14). "The Lord is righteous in all His ways, and holy in all His works" (Ps. cxlv. 17). That is,—not simply that His throne is just, but that it lives in justice; that is its object: not simply that His ways are holy, but that He is holy in having any ways at all. His whole creative impulse is holiness. As Jeremiah expresses it,—"The Lord God is the truth" (x. 10, see *Hebrew*). Hence Elihu :—" If He set His heart on Himself:" that is, if He did the very thing that Dr. Hodge says He always does; if He had no care but for Himself; if He took to Himself His creative enterprises, Elihu tells us, He would drop motiveless: "All flesh would perish together, and man would return to the dust" (Job xxxiv. 14, 15).

We defer further quotations till chapter fifth. "God is love; and he that dwelleth in love dwelleth in God, and God in him" (1 Jo. iv. 16); and though this is not inconsistent with His justice, it is inconsistent with Dr. Hodge. He cannot be so loving as to choose that as the very expression of His nature, and yet be so philosophically the reverse as to do everything for a selfish end.

§ 5. *Argument from Reason.*

The favorite argument that God does everything for Himself is, that He is an infinite Being, and, therefore, that there can be no higher to whom He can devote His agency. This reasoning has warped to it other propositions of the ten. It is built upon the idea that *being* must be the object of what He does, and that there can be no such thing as *quality* to ride paramount over any considerations of its possessor. Hence the doctrine that follows,—that the will of God is the ground of moral obligation. This denies quality an independent worth. And when there follows the tenet, that the idea of God is innate, we might expect the form in which that innate idea is stated (Theol., vol. i. p. 199), viz., that it is a sense of responsibility. Put those three things together,—first, that God is the Great End; second, that He makes *right*, which, of course, clears away the danger that that quality would sue to Him as His chiefest object; and third, that the idea of Him is innate, and consists in a sense of His authority,—and, of course, we have a trio of thoughts highly calculated to lift Him up to an arbitrary place, and to make it plausible that He should be the chief end of all His doings.

But if He is holy, and that holiness is something virtuous in itself; if two and two are four, not by the will of God, but by necessity in the fact affirmed; if a cloud is beautiful, not by the Creator's will, but of necessity when once it has been made, then we are prepared to feel how paramount holiness is. If God sovereignly loves it; if it is supreme; if He loves it beyond any other thought, nay, bows to it; if it be absolutely imperative over all His ways; then who will say that He cares for a creation beside that great object, viz., His love, or that He dares to have one except in submission to His character.

Our great doctrine, therefore, is, that holiness is God's highest end. We argue for it, first, because it is His delight; second, because it is His rule; third, because it is impossible to conceive of God as having any higher; and if now any one argues,— That is the very point; God hath nothing higher; He loves nothing better than holiness itself; and this is where your point is seen to be identical with that of Dr. Hodge; God loves holiness, and will do anything in the world to gratify it; He loves it beyond any other claim; and, therefore, holiness as His chief end and Himself as His chief end are but different aspects of the same reality,—we reply just there with our most careful thought, and with such fixed distinctions as must serve us through all this reasoning.

God loves holiness supremely; then is God the end, or holiness the end? (1) God loves holiness supremely; then, in indulging holiness, does He not gratify God? Of all other living critics Dr. Hodge will not make holiness aim at gratification. God's gratification is one thing. God's great aim is another. The very purity of holiness *in foro humano* is its preciousness for itself. God cannot be chief end of His works simply because He delights in being holy; for then personal delight would be His motive, and would be superior to holiness. Therefore, a little deeper (2);—God loves holiness supremely, and, therefore, holiness in *uno aspectu* is His highest end. But God loves holiness in Himself, because He loves holiness most of all when it is infinite; and it is found in an infinite condition only in Himself. Moreover, He is its Prince and norm. There is no holiness except in His creation; and whatever we shall presently decide as to the ground of the moral trait, certainly the embodiment of holiness is only in the Most High.

This evidently is at the head of true thinking in the case; and if we start legitimately here we shall be led the readier through all the wanderings of the discussion.

"God only wise!" That is, indeed, undoubted. But, rigidly, what does He love? Does He love God, or love wisdom? Take wisdom in Solomon's sense, and what does God love? Does He love self? or does He love this Eternal Wisdom? Or, to be fairer in the question, take out of Himself His holiness, and what does He love most? His happiness and all that may be left? or the holiness that has been separated away?

We cannot, therefore, honor God, and we cannot adore holiness, which in Him is infinitely great, without we set it up as unspeakable in itself, and as gloriously beyond anything else that could be terminal with the Most High.

Because, witness our appeals. Suppose it were altercated, and pushed angrily to the last, that God *was* His chief end; how could we settle it? Could we argue through six pages without one form of appeal? How strange it would seem if six pages were written and no man stumbled upon the query, which form of end was *right*. Now, what does that mean? Certainly that right rules the day; that God could no more appeal against it than against the beauty of a flower; that He considers it in all He does; that He values it of all He is; that He sets it above all His aims; and that one jot or tittle of its demand shall not be lost, but all its last end fulfilled.

§ 6. *The Opposite, Fetichism.*

When a Hindoo lets go the divine, he is

obliged to accentuate the other part of his idol. He began, the old writings tell us, with a symbolled Deity. He gathered up a piece of clay or a piece of bark, and distinctly beginning with the idea that it was what it was, he simply conceived it as instinct with the Deity. Soon, however, abandoning the diviner traits, he was obliged to accentuate the clay and the wood. This seems to be the origin of all idolatry. Not liking to retain God in our knowledge, we feel foolish without something, and the brute grows into the Deity. So in these doctrines. The orthodox, being of the same blood as others, priding themselves against molten shapes, and being entirely cut off historically from a tolerance of outward rite, nestle in doctrine. Doctrine psychically cannot save us. Being our pride, and being the thing we keep attending to assiduously above other men, what is to hinder it that it be our Deity? Letting holiness escape, as it must do when it decays, and letting it go stealthily, as it has done out of the Deities of an older worship, why should not doctrine appear, just as we have shown it does, bereft of holiness, and why, when the diamonds have been taken out, should they not be replaced by paste; that is, why should we not accentuate what is left by bringing into it other ends, and by making God Himself more God through His sovereignty itself, and by ghostly forms of less reasonable devotion?

Remembering that God is what we make Him— that is, that God to us is what we conceive Him to be, if we make the wrong God He is not worthy of respect, and may be treated as a heathen Deity. If Dr. Hodge takes holiness out of God, he has sacrificed Him in that act. Afterward He is an idol. And if we were shrewd enough to say,—Now he will accentuate the rest: denying holiness to be God's highest end, and denying morals to have a foundation in themselves, and denying faith to have a moral essence, and denying helplessness to have its essence in the will, he will go on and crust heavier the attributes that are left; we would but be calculating as from the past; we would but be standing in our tower, and hearing of the weather as from other posts; we might predict,—We shall have cartilage for the ancient bone; we shall have sovereignty for God's appetite for the right; Self instead of Eternal Excellency; Will instead of Holiness; and when we writhe under the painting of the Pit, we shall have God choosing it for personal display, and not promising to effect by it the most glorious creation.

Let us not anticipate, however.

CHAPTER II.

THE WILL OF GOD THE GROUND OF MORAL OBLIGATION.

§ 1. *Dr. Hodge's Statement of the Doctrine.*

ON page 405 of the first volume of his Theology Dr. Hodge introduces the subject, "*The will of God as the Ground of Moral Obligation.*" He is very precise in settling its boundaries:—"The question on this subject is, whether things are right or wrong simply because God commands or forbids them? Or, does He command or forbid them because they are right or wrong for some other reason than His will?" He then denies that they are right because they tend to happiness, and also rejects the doctrine that they are right because they promote our own happiness. "Others, again," he says, "place the ground of moral obligation in the fitness of things. There is, they affirm, an eternal and necessary difference between right and wrong, to which God, it is said, is as much bound to be conformed as His rational creatures." He then goes on to say,—"The common doctrine of Christians on this subject is, that the will of God is the ultimate ground of moral obligation to all rational creatures. No higher reason can be assigned why anything is right than that God commands it." Then further,—"In all cases, so far as we are concerned, it is His will that binds us, and constitutes the difference between right and wrong." Many pages farther on (vol. ii. p. 127), that we may be fair to Dr. Hodge, and show his abiding bent, we have this language,—"Whatever He commands is good, and whatever He forbids is evil." (Right, no doubt; and not interfering with its being what it is solely in the nature of things.) "The question is determined by authority." (True, unquestionably; and one can have nothing to say against it.) But again:—"*We cannot answer it from the nature of things.*" That is,—If I take a child and torture him with knives, that it is wrong must be "determined by authority." I cannot know, "from the nature of things." This is more frightful in the third volume. On page 260 we read,—"The slightest analysis of our feelings is sufficient to show that moral obligation is the obligation to conform our character and conduct to the will of an infinitely perfect Being, who has the authority to make His will imperative, and who has the power and the right to punish disobedience. The sense of guilt especially resolves itself into a consciousness of being amenable to a moral governor. The moral law, therefore, is in its nature the revelation of the will of God so far as that will concerns the conduct of His creatures. It has no other authority, and no other sanction, than that which it derives from Him. The same is true with regard to the laws of men. They have no power or authority unless they have a moral foundation." (Most true! but now notice what that is.) "And if they have a moral basis, so that they bind the conscience, that basis must be the divine will. The authority of civil rulers, the rights of property, of marriage, and all other civil rights do not rest on abstractions, nor on general principles of expediency. *They might be disregarded without guilt* (!), were they not sustained by the authority of God. All moral obligation, therefore, resolves itself into the obligation of conformity to the will of God."

We have come to something vital, certainly.

That is, I might take a child and seat him on a red-hot stove, "and all [his] civil rights might be disregarded without guilt, were they not sustained by the authority of God."*

The third doctrine comes in here to help out Dr. Hodge, viz., that the Idea of God is Innate; or the horror of his scheme would stand out in all its ghastliness. Suppose I did not know God! But I must know Him if the idea is innate. See how a system props itself together! Still, suppose I know Him dimly. Suppose I am the worst man that ever lived, but have the dimmest notion of the Most High, then all my acts are innocent as the slumbers of a child, except so far as they impinge upon the authority of God; and my guilt thereanent, according to principles that Dr. Hodge has confessed (see vol. i. pp. 195, 196), must be sharply graded by

* The neglect of old authorities on this point seems one of the wonders of Dr. Hodge's position. Listen to Bishop Butler:—"However, I am far from intending to deny, that the will of God is determined by what is fit, by the right and reason of the case; though one chooses to decline matters of such abstract speculation, and to speak with caution when one does speak of them. But if it be intelligible to say, that *it is fit and reasonable for every one to consult his own happiness*, then *fitness of action, or the right and reason of the case*, is an intelligible manner of speaking. And it seems as inconceivable, to suppose God to approve one course of action, or one end, preferably to another, which yet His acting at all from design implies that He does, without supposing somewhat prior in that end, to be the ground of the preference; as to suppose him to discern an abstract proposition to be true, without supposing somewhat prior to it to be the ground of the discernment. It doth not, therefore, appear, that moral right is any more relative to perception than abstract truth is, or that it is any more improper to speak of the fitness and rightness of actions and ends, as founded in the nature of things, than to speak of abstract truth as thus founded."—*Butler's Analogy (Harper's Ed.)*, page 173.

my knowledge of the Almighty. All guilt must depend upon God's being an innate idea; for if He were an acquired idea, the man not acquiring Him could not sin. And if He be an innate idea, the degree of its inborn-ness must regulate my guilt; and I may act a very incarnate fiend, and yet be as innocent as Gabriel, but for the government of Heaven.

Now think of this. All virtue in myself; all thought of wrong as wrong, apart from the will of the Almighty; all pity to the poor till I have thought of it as acceptable on high; all cruelty to that child, or shudder at his shrieks as soul-withering and sin-manifesting in their very selves; all right quality of act, outside of obedience to a King, is made utterly impossible. And the whole is followed by that remorseless speech,—that it is "*the common doctrine of Christians on [the] subject*" * (vol. i. p. 406).

§ 2. *Dr. Hodge's Contradictions.*

And yet, with more than common speediness,

---

* And yet the opposite was found plausible enough to win entrance from Dr. Hodge into the "Repertory" a third of a century ago. It is in one of the untraced articles:—" When the author, in his first chapter, makes 'the will of God' the only foundation of moral obligation, of course we understand him to mean that the distinction between moral good and evil is not arbitrary, or might have been the very reverse of what it now is, if God had so willed it, but as maintaining that the will of God, as His nature, is immutably inclined to good. As there is an extreme opinion on this subject of the will of God being the ultimate standard of moral rectitude, it would have been well to guard against this by an explanatory clause." *Bib. Repertory, Rev. of Junkin on Justif: Apr.* 1840, p. 271.

Dr. Hodge specially contradicts. Let us look at this with the utmost effort at distinctness.

Recollect (vol. i. p. 406), he has quoted our doctrine, and most explicitly refused it:—" Others, again, place the ground of moral obligation in the fitness of things,* which they exalt above God. There is, they affirm, an eternal and necessary difference between right and wrong, to which God, it is said, is as much bound" (we would say,—more bound) "to be conformed as are His rational creatures."

Here, then, is the proof that the *per contra* was not out of the mind of Dr. Hodge, but was weighed in his reckoning, and was distinctly meant to be denied.

And yet mark his progress. Hardly has he uttered his doctrine (p. 406) as to "the ultimate ground of obligation ; " hardly has he enforced it :— " No higher reason can be assigned why anything is right than that God commands it ; " hardly has he diverted it a little, because "*ground*" is one thing and "*rule*" is a very different thing, and he says,— "This means (1) that the divine will is the only *rule*," etc.; hardly has he set his doctrine in broad and very emphatic array before us, before he entirely and most astonishingly denies it :—"*By the word 'will' is not meant any arbitrary purpose,*

---

* A very imperfect expression. We would rather say, "*in itself.*" Moral is moral in itself considered ; or holiness finds its obligation in its intrinsic nature. "The fitness of things" is a poor periphrastic definition ; still, as no definition is ever precise, we accept it for what it obviously intends.

so that it were conceivable that God should will right to be wrong, or wrong right."

"Right to be wrong, or wrong, right!" Think of that! What would a man like Mill or Hamilton do with a section like this ninth? "The ultimate ground of moral obligation is the will of God." We are not to "place the ground of moral obligation in the fitness of things," or "affirm an eternal and necessary difference between right and wrong." "No higher reason can be assigned why anything is right than that God commands it." And yet the moment He is lifted to the throne Dr. Hodge shrinks quite away. There is a right and there is a wrong that He will never meddle with. What right? and what wrong? If the ground of moral obligation is the will of God, and but for His will in law we might utterly disregard it, why strip Him of His eternal holiness, and sink all grand distinctions, and make them mere choices of His will, and then mock Him in the end by the ghost of His dishonored excellencies?

And look further still (p. 406):—"The ground of moral obligation is the will of God. No higher reason can be assigned why anything is right than that God commands it." And, yet, a few sentences further:—"SOMETIMES things are right simply because God has commanded them; as circumcision," etc. Now put all together! First, there is no "eternal and necessary difference between right and wrong." Second, "things are right or wrong simply because God commands or forbids them" (p. 405).

Third, "He does not will right to be wrong, or wrong right." And, fourth, "*Sometimes* things are right simply because God has commanded them."

Like a serpent, Reason stings the heel that treads upon it.

We do not deny that there were plausibilities in the mind of Dr. Hodge. We do not conceal the sentences that lie between these unassimilable things. We mean to take those sentences and discourse of them and make them the basis of our assault. But demonstrably these four cannot tally. To say, "White is white," and then a sentence afterward, "Black is black;" and then after whole volumes, "Black is white,"—may find those volumes eloquent and able as one may please, and yet a scratch of the pen ought to be sufficient to show the error.

§ 3. *Argument from Reason.*

We pass over all texts from Scripture because Dr. Hodge resorts to none. If we quoted any, it would be those in which God is put quite on a plat with us. We are said to be in His image (Gen. i. 27). This is indeed the first dogma of Scripture ethics. We are bid afterwards to be "holy because [He] is holy" (Lev. xix. 2); and after that to be "holy as He is holy" (1 Pet. i. 15). In the Psalms of David we are told that "the righteous Lord loveth righteousness" (Ps. xi. 7); and in the Epistles of John (speaking of a fundamental principle of morals), "which thing is true in Him and in you" (1 Jo. ii. 8).

Now in all these respects how are we like Him? and where do we find our motive? and how are we holy as He is holy? and how can He be said to love righteousness? and how can a true love be true in Him as well as in us? when the very essence of right is manufactured by a decision of His will, and when *ex vi* there is none till He has made it, or declared what it shall be.

Dr. Hodge shrinks from all this, and takes refuge almost immediately in expressions about the divine "*nature.*" These fill up the interval. These bridge the chasm between the opposites that we have just been quoting. "The will of God," he says, "is the expression or revelation of His nature" (p. 406). Would it be unfair to demand that Dr. Hodge shall choose? Or, excusing him from that, would it be unfair to say, He certainly teaches one of three things,—either, first, that the will of God is the ground of moral obligation, or, second, that the nature of God is the ground of moral obligation, or, third, that the two are one thing; for that the nature and the will are the same in all judgment of the case? He certainly says the second,—"So that the ultimate foundation of moral obligation is the nature of God" (p. 406). We have seen that he says the first. So that our most charitable supposition is to infer from him the third, and to suppose that he prefers that for his final statement.

Now if the ultimate ground is both Will and Nature, it is fair to ask whether it would be right in God to be different in both of these. If the ground of right's being right is to be found in the existing Nature, and Dr. Hodge distinctly denies that it is right from a nature in itself, then change the Nature, and we change our existing holiness. That is no forlorn supposition. In *our* theory God's Nature cannot be changed, because it is intrinsically holy. We speak of an "infinite goodness," and an "immutable excellence," and so does Dr. Hodge, strangest of all, on this same page (406). *We* believe it, however, while Dr. Hodge distinctly declares that there is no "eternal difference between right and wrong." Now, if there is no "eternal difference between right and wrong," we can *suppose* at least that God's Nature had been opposite, and then it is fair to ask,—Would all moral distinctions have been actually reversed?

The supposition is impossible.

The supposition is impossible that anything should be different from what it is. And yet the Bible makes just such suppositions (Ps. lxxxi. 13, 14; Lu. xix. 42). They let in behind the fact a distinct light upon its reality. If God's Nature were opposite it would be wicked; but why wicked if it is His Nature? The theory incontestably breaks down. God can make wrong right, or right wrong; or else there is something as eternal as Himself, that "He got possession of as the beginning of His way" (Prov. viii. 22*).

Again, suppose I *resist* the will of God. You say, I will be punished. But that does not make it

* See *Commentary*, in loco.

wicked. You say, I will be guilty. But what distinctly do you mean by that? If you mean, I shall be exposed to punishment, I might say, I will endure that. What is it that obliges me to obey the will of God? It is impossible not at last to be obliged to say, Because it is *wrong* not to; so in the *dernier resort* right prevails above will, and we are obliged to underpin even the will of the Almighty by the consideration that it is wrong to violate it.

§ 4. *The Error Fetich.*

And yet in a system that gets rid of intrinsic holiness, the emphasizing of will is the instinct of nature. Losing the kernel we accent the shell. This is true of all these AUTHORITY-DOCTRINES. Stealthily suffering old-fashioned rightness and wrongness to drop out of our faith, to thicken up the rest is a matter of necessity. Hence a complete system of *natural* traits: authority instead of excellence; will instead of character; responsibility instead of conscience; a God not humane even to the extent that man is, but supreme and regal, and offering a retreat for this in what is now to be called an innate idea.

CHAPTER III.
THE IDEA OF GOD INNATE.

§ 1. *Dr. Hodge's Statement of the Doctrine.*

IN Vol. I., page 191, of his Theology, Dr. Hodge distinctly announces that "*The Knowledge of God is Innate.*

Our doctrine is that there is an eternal holiness, that we get the idea within, that we have certain emotions, that those emotions are consciously right just as other emotions are of conscious beauty, that right is the main attribute of God, that God is an inferred idea like heaven or like other spirits than ourselves, that all His attributes are inferred from attributes that are discerned in man, and, therefore, though a simple being, He is a most complex idea, in no part innate, but in all parts inferred and combined, and that when we have given Him the attribute of right, we have given Him the critical trait, and all else—the power and the infinitude—may be more safely trusted to frame themselves together.

Dr. Hodge, making the Deity innate, gives us the precision of knowing that he cuts himself off intelligently from this induction. He says (vol. i. p. 199), "The knowledge of God is not due to a process of reasoning." He, like us, admits a revelation. He is not speaking of that source of divine knowledge that we have in Scripture. He is speaking of that which comes to us besides. And while we trace it to an induction from ourselves, he, of course, cannot. He discards an intrinsic holiness. He builds his moral distinctions upon the will of a Supreme. And, therefore, he has pushed himself off from inference. Like all other fetich he must resort to some wizard thing; and, therefore, the thought of something innate comes admirably into play.

Look at it all combined.

(1) There is no holiness as the final aim but a God all for Himself. (2) There is no holiness that is holy in itself, but a holiness that is the will of the Almighty. And as that separates Him from our image, and takes away from Him motives of right, (3) there is no God that we can find pictured from ourselves; no Deity that can be learned, lifted into light by making infinite attributes of our own; but there is an innate Deity, and Dr. Hodge distinctly tells us what, viz., a Deity corresponding to his previous account,—a God all for Himself, and supreme even over right, that is, as he now expresses it, a God innate in our hearts, and innate in just two ideas, that is, of RESPONSIBILITY and DEPENDENCE.

"Men have the conviction that there is a Being on whom they are dependent, and to whom they are responsible." This conviction is opposed "to that acquired by a process of research and reasoning" (vol. i. p. 191). "[God's] existence is a self-evident truth" (p. 23). "It is in the general sense of a Being on whom we are dependent and to whom we are responsible." "If this idea is analyzed it will be found to embrace the conviction that God is a person, and that He possesses moral attributes, and acts as a moral governor." "All that is maintained is that this sense of dependence and accountability to a being higher than themselves exists in the minds of all men" (vol. i. p. 195).

§ 2. *Dr. Hodge's Contradictions.*

But hardly has all this escaped from the lips of

Dr. Hodge before he begins most singularly to contradict it. On page 339 he asks,—"*How do we know God?*" He approaches this systematically. He uses the very word, "*idea.*" "How does the mind proceed in forming its idea of God?" He shows distinctly how it can be formed, and depends to form it upon the very methods that he had once disowned. He says,—"We deny to God any limitation; we ascribe to Him every excellence in the highest degree; and we refer to Him as the great First Cause every attribute manifested in His works. We are the children of God, and, therefore, we are like Him. We are, therefore, authorized to ascribe to Him all the attributes of our own nature as rational creatures, without limitation, and to an infinite degree. If we are like God, God is like us. *This is the fundamental principle of all religion.* If we are His children, He is our Father, whose image we bear, and of whose nature we partake. This, in the proper sense of the word, is Anthropomorphism, a word much abused, and often used in a bad sense to express the idea that God is altogether such a one as ourselves, a being of like limitations and passions. *In the sense, however, just explained, it expresses the doctrine of the Church, and of the great mass of mankind.* Jacobi well says: 'We confess, therefore, to an Anthropomorphism inseparable from the conviction that man bears the image of God, and maintain that besides this Anthropomorphism, which has always been called Theism, is nothing but atheism and *fetichism.*'"

Now, to say the very least, is not the doctrine of Dr. Hodge "*besides* this Anthropomorphism?" Nay, is it not against it? I have conscience. I have a sense of right. I have a love of truth, on its own account. · Dr. Hodge has articulately stated that God, my image-bearer, has nothing of the kind. I have holiness as my highest end, and yet Dr. Hodge takes pains to say that God's highest end is but Himself; and, moreover, that there is nothing right in a sense that he specifically states "in the fitness of things," or, as he boldly puts it forth, through "an eternal and necessary difference between right and wrong."

There are, therefore, two impinging contradictions in Dr. Hodge. The first is a historical one. We do not get our idea of God innately if we get it by putting our own attributes together. And, secondly, if we put our own attributes together we would not frame Dr. Hodge's God, for his God discards for attributes what is intrinsical like ours, and implants as an idea what is a mere responsibility to His will.

§ 3. *Argument from Scripture.*

By a singular fatality Dr. Hodge rests the whole weight of this upon one passage. He leans upon that one remarkably, and with a singleness that it is hard to explain. "The Bible asserts," he says, "that the knowledge of God is thus universal. This it does both directly and by necessary implication. The Apostle directly asserts in regard to the

CHAP. III.]   *The Idea of God Innate.*   115

heathen as such without limitation, that they have the knowledge of God, and such knowledge as to render their impiety and immorality inexcusable." " He says of the most depraved of men, that they knew the righteous judgment of God, that those who commit sin are worthy of death" (Rom. i. 32). " All this is done without any preliminary demonstration of the being of God. It assumes that men know that there is a God, and that they are subject to this moral government." And he quotes, to substantiate all this, Rom. i. 19–21. " Because that when they knew God, etc." I say, he quotes it with a singular fatality, because this passage, of all other passages in the Bible, places this knowledge exactly where Dr. Hodge denies it to be, viz., on the plat of observation and experience! Do not let us look carelessly at this. It is vital to our view of the subject, and awkward, to say the very least, as the selected testimony of Dr. Hodge (see vol. i. p. 195).

The Apostle is indeed proving the extensiveness of the knowledge of God. Whether he pronounces it universal I can hardly say. He is speaking " of men who hold the truth in unrighteousness" (v. 18). But instead of saying that they get their knowledge as innate, he expressly says the contrary. He says, " God hath shewed it unto them" (v. 19). He tells us *how* He " hath shewed it unto them." " For the invisible things of Him from the creation of the world are clearly seen, being understood by the things that are made, even His eternal power and Godhead, so that they are without excuse." He does go on to say,—"Because that when they knew God;" but it is superfluous to add, that he has given us the only mode in which any man has a right to quote him as an authority in the case.

"The heavens declare the glory of God." We believe that the Divinity is specifically *not* innate. We believe that the idea is laboriously and empirically acquired ; or, that it has been revealed ; that it is hereditary ; that it may be increased ; that it must be studied, and kept accurate by the patience of the saints ; that it would be as right to say that a logarithm was innate, as that God was an innate idea ; and, moreover (with humility and respect), that such an idea ought not to have been conceived, inasmuch as God is one of the most complex notions of the mind, and only simple things are conceived of as innate even by that school of (as we believe) mistaken metaphysics.

§ 4. *Argument from Reason.*

Dr. Hodge argues for it, however; and attempts the usual tests of *universality* and *necessity.*

But let us quote his language :—

"The question now is, whether the existence of God is an intuitive truth? Is it given in the very constitution of our nature? Is it one of those truths that reveal themselves to every human mind, and to which the mind is forced to assent? In other words, has it the characteristics of universality and necessity?" (i. p. 194.)

Let us be very rigid here. These are the very essences of Dr. Hodge's Books.

He goes on:

"It should be remarked that when universality is made a criterion of intuitive truths, it is intended to apply to those truths only which have their foundation or evidence in the constitution of our nature."*

Let us pause a moment.

The idea of a God has its foundation or evidence in the constitution of our nature. Things that have their foundation or evidence in the constitution of our nature are to be known by universality and necessity. Universality can only be pleaded in the instance of those truths "which have their foundation or evidence in the constitution of our nature!" (i. p. 194.)

There is something loose there certainly.

But let us go on.

"When it is asked whether the existence of God is an intuitive truth, the question is equivalent to asking, Whether the belief in His existence is universal and necessary? If it be true that all men do believe there is a God, and that no man can possibly disbelieve His existence, then His existence is an intuitive truth. It is one of those given in the constitution of our nature; or which, our nature

* What drives Dr. Hodge to this is remembering that mistakes have been universal, as, for example, the rising of the sun. "If ignorance be universal, error may be universal. All men, for example, for ages believed that the sun moves round the earth; but the universality of that belief was no evidence of its truth" (i. p. 194).

being what it is, no man can fail to know and to acknowledge" (i. p. 194).

After giving up universality as universality, and going over for what it was to prove largely to the thing itself, it would be odd if Dr. Hodge should give up universality as absolutely universal, and be driven in distress in that direction, too, to some other test.

And yet, listen to him.

"Even if the fact be admitted, that such tribes have no idea of God, it would not be conclusive (!). Should a tribe of idiots be discovered, it would not prove that reason is not an attribute of our nature" (p. 196). But why? Because,—"It is hardly conceivable that a human soul should exist in any state of development, without a sense of responsibility, [i. e., the very thing to be demonstrated!] and this involves the idea of God."

How singular this thing is,—Dr. Hodge, one of the ablest of our Church; keen in argument; noted as a veteran in debate; and lucid and emphatic beyond almost any one beside; and yet, in chase of a most dangerous thought, doubling upon his track in a way unconscious to himself! The idea of God is innate (vol. i. p. 191). An idea is certainly innate when it is universal and necessary (p. 194). Nevertheless, an idea may be universal and yet not be innate, as, for example, the rising of the sun. An idea innate, because universal, must be an idea universal and also having its foundation in the nature of things (p. 194). Moreover, the knowledge of

God is said not to be universal (p. 196). Grant it is not (!). That would not prove it not to be innate (p. 196). Nevertheless, it *must* be universal; because how could a human soul "exist without a sense of responsibility, and this involves the idea of God?" (i. p. 197.)

Let no one say,—This is *ex facie* caricature. I beg that the pages may be searched in the same fair way in which they have been articulately quoted.

But the above is not all. *Our* principle is, that there is a *per se* and natural holiness. Dr. Hodge denies it. We hold that there is an intrinsic difference between right and wrong, and that God Himself accepts it. Dr. Hodge states this doctrine, and ranges it with those that are palpably erroneous (i. p. 406). He comes now to his idea of God, and cannot make it, as we can, out of a conscious knowledge of the emotion of holiness. He, therefore, pronounces it innate. He proves it innate, because it is *universal* and *necessary*. He stumbles, we have seen, under the first; but then we expect he will make it up when he comes to the second. What is our amazement, when he has faltered visibly as to the *universal*, and thrown all the weight upon the *necessary*, to find him appealing, as his main proof that it is *necessary*, to the fact that it is *universal!!!* We scarce know of such a thing in theological literature. "If it be admitted that the knowledge of God is universal among men, is it also a necessary belief? Is it impossible for the mind to dispossess itself of the conviction that there is a God? Necessity, as remarked above, may be considered as involved in universality, at least in such a case as this. There is no satisfactory way of accounting for the universal belief in the existence of God, except that such belief is founded on the very constitution of our nature" (p. 197).

I said "main proof." If any one objects to that, and says I ought to have been satisfied with saying, it is his first proof, I will give the other most gladly. It is, that men implicitly believe in their moral nature. We have been arguing for just that thing with all our might. We have held that there is an intrinsic holiness; that men are conscious of it; that it is a dictate of conscience; that it is inwrought in the very constitution of our being. Dr. Hodge is also aware of the like, for he taxes some people with it and goes against it as destroying the sovereignty of Heaven (p. 406). And yet from words and phrases that are certainly more ours than his he lays a foundation—for doing what? Now let us be very precise and pains-taking. We need the very edge of what is definite. He lays the foundation of doing what? Does he claim our doctrine? That would be strange enough. But he claims it for the very purpose of defeating it. He comes over to our side, and takes a doctrine which looks right into our point of intrinsic holiness; which he himself could be forced to admit was, at the very least, as favorable to us as it could possibly be dreamed to be to him,—and makes it the foundation of overturning our ground, and building up his own.

He does it first under the former head. When driven into difficulty about the *universal*, and forced to look out of itself for its own support (though let it be remembered, it was to be itself a support, and an intrinsic argument) he turns at once to our position. " Unless such people show that they have no sense of right and wrong, there is no evidence that they have no knowledge of such a being as God" (i. p. 197).

This he pursues *ad unguem* under the head of *necessity*. " The simple fact of Scripture and experience is, that the moral law, as written upon the heart, is indelible" (p. 198). That we believe, and connect it with the consciousness that the moral law is excellent in itself. Dr. Hodge accepts it also; inconsistently we see, but of that we are not now to complain. The point we wonder at is that he should take this as his argument. *Universality* had half failed. *Necessity* seemed now everything. And after having cleared the track, and shown that there are three kinds of necessity; first, of truisms, such as that a given part is less than the whole; second, of externals, as that "a man cannot deny that he has a body;" and, third, of such facts as the existence of God,—we might naturally expect the very firmest and most conclusive reasoning. Instead of that we have nothing but this premise common to both.

That man has a moral nature, in that he sees a difference between right and wrong, it was natural to suppose that Dr. Hodge would declare could be *reconciled* to his teaching, but that he should adopt it as his chiefest argument; nay, make it the only one, as in its very self the same as a sense of responsibility to God (p. 197); and that he should do this without apparent consciousness that in the adverse theory it was to say the very least equally in place,—is to take one of the grandest arches of his system, and to seek for it an abutment such that we can look under it, and actually see, that it can have no separate foundation.

§ 5. *The Idea conceived of, a Fetich.*

That the idea of God is innate, if Dr. Hodge had been satisfied with that, might have seemed to be a harmless error. But he has gone further. He has told us what that innate idea is.

Now, sadly enough, it is one of those Godnesses, if I might coin such a word, which is no God at all, any more than the metal of an idol. He tells us that the idea we have is of Responsibility to a Supreme.

Now I boldly aver that there is not in this the least particle of worship.

Let me be clear.

Had God shown Himself, then I confess there would come responsibility. Give me my idea of God; that is, of a being of intrinsic goodness; make Him all I care for, that is, all I feel endangered about, I mean a thought in my mind of ineffable holiness, and then I become responsible. But Dr. Hodge sets responsibility at the top. That is he makes God all for Himself; he makes God absolutely manufacture holiness; he makes goodness mean obsequiousness to His will; and then, of course, he makes my very idea of God consist in responsibility. Now, boldly, such responsibility is not even good.

This is just what the idolater does. He expels holiness from his idol, and then worships him out of a responsibility which is a dream.

Our argument, therefore, is twofold. If feeling responsible were a duty it would not be my grand duty, nor would it be the grandest way originally to express it. My grandest duty is to holiness. God is nothing to me; I am not bound to worship Him; I do not care for Him, till I find out that He is holy. That He created me is no claim at all. He wronged me, unless He can be holy. My heart, which He has created in His image, goes searching about till it can think of Him as holy. When that happens I can feel that I am responsible.

One sees, therefore, our two points. We cannot feel responsible till we think of Him as holy, and, therefore, a sense of responsibility *semet ipso* is a cheat, first, because the right responsibility must have known of Him *a parte ante*, and second, because the wrong responsibility can be no formed idea of Divinity at all.

CHAPTER IV.
VINDICATORY JUSTICE AS A PRIMORDIAL ATTRIBUTE OF GOD.

§ 1. *Dr. Hodge's Statement of the Doctrine.*

By an odd coincidence, when I had written this

heading and felt too fatigued to go forward with the work, my eye fell upon a copy of Ossian, covered with dust, that had been thrown upon a chair near me; I opened it hap-hazard, and came upon this as my very first sentence :—

" 'Annir,' said Starno of lakes, ' was a fire that consumed of old. He poured death from his eyes along the striving fields. His joy was in the fall of men. Blood to him was a summer stream, that brings joy to withered vales, from its own mossy rock.'"

It is this blood-thirsty character that we charge Dr. Hodge with implying in the Almighty.

To be perfectly fair, we will assume all responsibility by describing our creed first.

We believe that holiness is a quality: that holiness in its other sense, viz., of those things in which this quality is found, is two things, and that these two are both emotions :—first, a love to the welfare of other beings, and second, a love to the quality of holiness itself. These two are primordial. These two are all that are primordial; and all other emotions or acts are right or wrong, not simply as agreeing with these two, but as being, philosophically stated, but instances under them. Nothing, therefore, not benevolence, or not a love of right as intrinsically excellent, can be a primordial morality of God.

But God, in being benevolent, and in being regardful of holiness, finds punishment, in the very nature of the world, and by the very constitution that He has made, a good instrument for advancing holiness. It is so *naturaliter;* it is so from the very depths of our being.

It is not so for its effect upon its victim; for sometimes its victim may be accursed, as in the instance of the reprobate. But it is so over the sum of creatures. Punishment, therefore, is an instrument. A resort to it is not out of a primary trait, but out of a wide expediency.

Vindicatory justice, therefore, is not a primordial attribute in God, and scarce a secondary one. It is a bundle: I mean, a convenient phrase for putting together a whole story as to government. It means that God has two traits, benevolence and a love of holiness; that these two traits govern His Providence, and make Him desire holiness and desire happiness in all His creatures; that punishment is a means to promote them; that this means is in the very nature of the case; that He feels bound, therefore, to resort to it; that He has sworn, therefore, to resort to it; and as it is a means imbedded in the very nature of truth, there is no fear but that He will always feel bound for its administration.

Having sworn to punish sin, He is bound to. Having sworn to punish sin, He was, that shows, otherwise bound to. He is, therefore, bound in both ways, consequentially by His oath, and antecedently by the facts that made Him swear it. But they are not moral facts. The only intrinsically moral facts are the duty of benevolence and the duty of love to the principle of holiness. Punishment is a mere instrument. To fear it is mere nature, and to employ it is mere inferential duty; and in punishment itself there is no more sphere of morals, than in the axe or the froe, that a man may wield for an industrious maintenance. When, therefore, Dr. Hodge declares (i. p. 418) that a man has a conscious guilt, such that he feels that he has a desert of punishment, he is journeying into the intuitive with that which has about as much right to any such genesis for its belief, as our belief in food, or in any of the material dependancies of our being.

We believe God is just. But we believe that that means that He loves holiness; and, something more than that, that He knows how punishment can promote it; and, therefore, as an instance of His holiness, He is just, because He draws not back from that painful means which wisdom and holiness by necessity must approve.

Dr. Hodge, on the contrary, ranges benevolence and justice together, and makes them primordial alike. Let us listen to him. He says,—"We ascribe intelligence, knowledge, power, holiness, goodness and truth to God. On the same grounds we ascribe to God justice." There is much to object to in this:—for example, " intelligence " and "knowledge" mean the same thing; "goodness" is a part of "holiness;" and "truth" or truthfulness is not primordial any more than "justice," Paul himself finding a different generalization for it in his epistle to the Romans (Rom. xiii. 9). But what we are concerned about at present is the generalization of "justice." Dr. Hodge makes it a genus by itself. "[Men] also know intuitively," he says, "that God is just as well as holy; and therefore, that His moral perfection calls for the punishment of sin by the same necessity by which He disapproves and hates it" (i. p. 421). Satisfaction, he declares, is a thing demanded in itself considered. "A man, when thus convinced of sin, sees that not only would it be right that he should be punished, but that the justice or moral excellence of God demands his punishment. It is not that he ought to suffer for the good of others, or to sustain the moral government of God, but that he as a sinner, and for his sins, ought to suffer. Were he the only creature in the universe, this conviction would be the same both in nature and degree" (i. p. 421). Now what does this mean? Not simply that it is wise that he should suffer; for wisdom means skill in adapting instrumentalities to an end; but that it is right that he should suffer, in such a sense that there is a justice that demands it *semet ipso*,—a guilt that intuitively accepts it, and, therefore, a primary attribute that desires it, just as benevolence and its sister trait desire our welfare and desire the holiness of all the universe. One desire is just co-ordinate with the other two.

And, therefore, Dr. Hodge proceeds to assert the distinctness of justice and benevolence. He does not bring in the other trait, and therefore, I think he is unfair in his polemic. He does not speak of THE ADVANCEMENT of HOLINESS. He is always aiming at the "greatest-happiness-theory." He never speaks of the greatest holiness. He pits justice against benevolence. He would imply that that is the only alternative. And when he speaks against the Apostolic Fathers he makes Clemens Alexandrinus say,—" Men should remember that punishment is for the good of the offender and for the prevention of evil" (Paedagogus i. viii.), and then goes on to speak as though "evil" could only be, as opposed to happiness (i. p. 419). This is a vice all through these volumes. Yet still, as justice *is* the handmaid of benevolence, its serving a higher end does not prevent it from being misstated when it is declared to be distinct from benevolence. Witness, therefore, the statement on another page,—"If justice and benevolence are distinct in us they are distinct in God" (i. p. 420; see also 422). No one will challenge us, then, for the statement that Dr. Hodge's doctrine of justice is,—That it is on the same plat with benevolence, and hungers after its end, just as benevolence does after the welfare of the creation.

§ 2. *Dr. Hodge's Contradiction.**

Benevolence, though, is so primordial that if justice hungers after its end as directly as benevo-

---

* Sometimes we think of the contradictions of Dr. Hodge (even when elaborate) as wholly inadvertent, and as expressing no positive idea of Dr. Hodge whatever. On what possible theory can we reconcile the four hundred and ninety-seventh and the four hundred and ninety-ninth pages of the third volume? — On page 497 we read,—" The victims then offered, having no inherent dignity or

lence, Dr. Hodge feels himself in unpleasant nearness to the doctrine of revenge. There begin to blow upon him the hot winds of a resentful anger; and he dreads ascribing to God what would be thoroughly disgraceful to a perfect creature. He takes the word "*vindictive*," therefore, and tries to make a distinction; just as he might take the word *selfish* under the first doctrine, and say,—God does everything for Himself yet God is not *selfish*. We see at a glance that the discrepance aimed to be expressed is a discrepance *verecundiæ*. *Selfishness* is used as a reproach. *Vindictiveness* is used as a reproach. *Vindicatoriness* means articulately the same. That *vindicatoriness* is not *vindictive* means simply that God may *vindicate*, and yet not be *wrong:* except, now, what we wish most carefully to notice, that vindicatoriness in a proper

worth, could not take away sin:" on page 499,—"The common doctrine as to these sin offerings is, (1) That the design of such offerings was to propitiate God; to satisfy His justice, and to render it consistent and proper that the offence for which they were offered should be forgiven; (2) That this propitiation of God was secured by the expiation of guilt; by such an offering as covered sin, so that it did not appear before Him as demanding punishment; (3) That this expiation was effected by vicarious punishment; the victim being substituted for the offender, bearing his guilt, and suffering the penalty which he had incurred; (4) That the effect of such sin offerings was the pardon of the offender, and his restoration to favor and to the enjoyment of the privileges which he had forfeited. If this be the true Scriptural idea of a sacrifice for sin, then do the Scriptures in declaring that Christ was a sacrifice, intend to teach that He was the substitute for sinners; that He bore their guilt and suffered the penalty of the law in their stead."

sense may be most hopefully distinguished from all that is vindictive. Vindicatoriness, in a proper sense, is just what we mean by justice, viz., that goodness and sanctity in God which lead Him to defend both; that benevolence which leads Him to vindicate our good, and that holiness which leads Him to vindicate our purity; that love of the quality of right which rules in God more than any other desire, and which leads Him to vindicate the law as the means of advancing the happiness and holiness of all His creatures.

Dr. Hodge, denying this, leaves no difference between vindicatoriness and vindictiveness: indeed, wash the odium from the latter, and in denying one, he denies all thinkable space, and all logical possibility, for conceiving of the other.

But that is not all he does.

He comes once and again to our own ground. On page 27 (vol. i.) he says,—" It is more congenial with the nature of God to bless than to curse, to save than to destroy. The Bible everywhere teaches that God delighteth not in the death of the wicked." And then in vol. ii. (pp. 204, 205),—" The infliction of suffering to gratify malice and revenge is of course a crime. *To inflict it for the attainment of some right and desirable end may be not only just but benevolent.* Is not the support of the divine law such an end?" Now what does this mean? Certainly not the support of God. It must mean "the support of the law" in the mind of the creature. And what is meant by the support of law in the mind of the creature? Certainly the encouragement of his character. It means the depression of the creature's sins, the advancement of the creature's holiness, and, however it may happen, whether in the victim or the race, the advancement of good in the whole intelligent creation.

§ 3. *The Scriptures that Dr. Hodge quotes.*

The Scriptures that Dr. Hodge quotes are of that common kind that cannot possibly be distinctive in the controversy. He says on the 416 page (v. i.). "The Bible constantly represents God as a righteous ruler and a just judge. These two aspects of His character are not carefully distinguished. We have the assurance which runs through the Scriptures that 'The judge of all the earth' must 'do right' (Gen. xviii. 25): 'God is a righteous judge' (Ps. vii. 11, *Marg.*). 'He shall judge the world with righteousness' (Ps. xcvi. 13): 'Clouds and darkness are round about Him'" (Ps. xcvii. 2). It is such texts that are quoted in all the argument. Now if any one were denying that God was just, these would be the texts; but as the point denied is that there are more than two tables in the law, these are not the texts to show that justice has a separate claim, and stands primordially among the attributes of the Most High.

On the contrary, there are Scriptures that prove that it has not:—

§ 4. *Scriptures that Refute the Error.*

In the XVIIIth chapter of the prophecy of Ezekiel Jehovah disclaims again and again that He has any "pleasure in the death of him that dieth" (see also v. 32 and chap. xxxiii. 11). Jeremiah repeats the assertion:—"For he doth not afflict willingly nor grieve the children of men. To crush under His feet all the prisoners of the earth, the Lord approveth not" (Lam. iii. 33, 34). Isaiah (xxviii. 21) speaks of His "strange work," and His "strange act." And Paul makes the doctrine definite,—"He would have all men to be saved, and to come to the knowledge of the truth" (1 Tim. ii. 4).

Now, of course, Dr. Hodge has a meaning ready for all these accordant passages, and it becomes us modestly to suppose that from the first. But still let me press the question,—Where is the *room* for any meaning? Among primordial traits where is the *chance* for such antagonism? If God be primordially benevolent, can He ever afflict willingly? And, therefore, if He be primordially just, must not His punishments satisfy a trait, and must not that trait *feel* satisfied, and not breed a conduct that is the object of its own aversion?

§ 5. *Argument from Reason.*

Dr. Hodge's great argument for his view of vindicatory justice is that man is conscious of a desert of evil. "We do not stop to ask, or to think, what may be the collateral effect on others of the infliction of punishment. Anterior to such reflection, and independent of it, is the intuitive perception, that sin should be punished for its own sake, or on account of its inherent ill desert" (i. p. 420).

It makes us helpless to have such things asserted. How can we appeal against any man's consciousness? And Dr. Hodge is full of this particular resort. A hundred times in his book he appeals to the consciousness of all men in all ages (see p. 229). Fortunately in this case he draws a practical inference. There we can meet him. He quotes, "Qualis homo, talis Deus:" "If any one knows himself, he will know God (Clemens Alexandrinus);" "The perfections of God are those of our own souls (Leibnitz)" (i. p. 374). Under the particular head of Justice he argues,—" Instinctive moral judgments are as clear and as trustworthy revelations of the nature of God as can possibly be made. If we in obedience to the nature which He has given us, intuitively perceive or judge that sin ought to be punished for its own sake, and irrespective of the good effect punishment may have on others, then such also is the judgment of God. This is the principle which underlies and determines all our ideas of the Supreme Being. *If moral perfection be not in Him what it is in us, then He is in us an unknown something, and we use words without meaning when we speak of Him as holy, just and good*" (i. p. 420).

We will not stop to seize upon these sentences as bearing upon the previous discussion. If "*Qualis homo, talis Deus*," how can God make everything for Himself? Christ speed the day when modern theology will make the Great Prince HUMANE in the better sense! We ought to learn the principle from the very mission of God Incarnate. But let that pass. What we wish specifically to press is that Dr. Hodge defeats himself. He could not have chosen a more fatal argument.

And let me beg now that there may be the most rigid reasoning; for here is the very essence of our common Christianity.

If God be "an unknown something" unless "moral perfection be in Him what it is in us," then man's justice and God's justice must be identically the same. Favorable to this precision is that anecdote which Dr. Hodge tells of an English judge who told a criminal,—"You are transported, not because you have stolen these goods, but that goods may not be stolen" (ii. p. 579). Dr. Hodge, therefore, seems to realize that what is asserted of God must in the end come home to man.

Now here we have a grand index. What justice is that which is in the bosom of man? Solely a remedial justice.

Not only can he not execute vengeance, but he cannot entertain it. We are to love our enemies. What is competent for an individual is competent for a court. Man personally and men in nations are identically the same. The judge who sits upon a bench weeps when he condemns a criminal. Why? Because the heart-claim and the law-claim are different: the heart-claim primordial,—the law-claim secondary. As a mystery, God's justice may be deified; but, if human, it has intelligible shape. Dr. Hodge, bound fast to his own simple dialectic, God's justice being man's justice, and every body knowing what man's justice is; everybody knowing that certainly it cannot take vengeance, and logically is remedial and makes a limit of what is beneficial in the State,—we need have no fear at all about the justice of the Almighty. If Dr. Hodge will only stand to that one principle of likeness to ourselves, I feel very sure that justice anywhere will never far or certainly never long harden into anger.

If that very language is appealed to,—God's "anger,"—or still stronger expressions, like God's revenge, and what is meant by "Vengeance is mine, I will repay, saith the Lord,"— I reply that of course every Scripture must have some strong Saxon sense; but, unless we are willing to be literal with all of them, and make God penitent (Gen. vi. 6) and "grieved" (Heb. iii. 10) and "furious" (Nah. i. 2) and "cruel" (Is. xiii. 9) and "weary" (Is. i. 14) and ignorant (Ps. liii. 2) and deceitful (1 Kings xxii. 23) and inquiring (Ps. xiv. 2), we must be satisfied to deal with each *in situ*. They are tropical. The trope is what it may prove to be in the context; and in the instance of revenge (Rom. xii. 19), it is but an assertion that God has the higher bench, and holds the final court for adjudication among men.

But Dr. Hodge will ask, How about the wrath upon the Redeemer? I would reply in passing, that Dr. Hodge shuts the door upon this with one of his strange contradictions. He says (vol. i. p. 423), —"If the prevention of crime were the primary end of punishment, then, if the punishment of the innocent, the execution, for example, of the wife and children of a murderer, would have a greater restraining influence than the punishment of the guilty murderer, their execution would be just. But this would shock the moral sense of men." It does no doubt. But so do many of the ways of the Most High. So with Achan. So with the children of the Canaanites. So with the sons of the drunkard: indeed, all the posterity of Adam. Is it not remarkable that a man who believes in direct imputation and in the vicarious suffering of Christ, should allow such a sentence to escape from beneath his pen in upholding one species of justice to the overthrow of another.

For look more specifically,—If I am thirsting for revenge, I can less afford to let one victim suffer instead of another. If justice is primordial, it whets its anger; it is not pleased with the murder of the innocent. If Christ was ever so much a volunteer, still, if God was righteous in taking vengeance in Dr. Hodge's sense, we cannot conceive that His own dear Son could satisfy His revenge. But if we were guilty, and that guilt was for the violation of law, and law itself was for the good of the universe, that good being of both kinds, natural good and moral good; punishment, therefore, being an expedient, and a stroke to which God was pledged,—substitution, naturally, wears just the look that is suited to such a case,—an artificial scheme adjusted to an artificial instrument. God's curse of Christ, if He really hungered to curse the wicked, is hardly thinkable: but that He should curse Christ under a covenant, punishment itself being under a sort of covenant; and that He should thereby satisfy the law, the law itself being a governmental expedient, —and thereby keep His oath,—Dr. Hodge himself admitting, as much as we do, that His atonement keeps it,—appears to us far more natural as a remedial revenge, than as a something of primordial wrath against the immediate sinner.

Let not Dr. Hodge say,—It is a mere drama. For, why? If sin really deserves punishment, and on punishment's account; if guilt really asks for it, and feels that it needs it, and that not as a reparation to its mischief in the creation,—then the cross is a real drama. Sin has not been punished. Guilt has not had its want. And the hunger of the Just has been fed on an opposite aliment. But if God loves holiness; if He loves it supremely beyond anything beside; if punishment is a means to lift it up; if Hell itself is to give it maintenance, and He has framed a law therefore, and sworn that it shall be enforced,—then He does not love Hell, but He loves holiness; and if He can keep His oath and yet spare the sinner, what He could not do if He loved vengeance on its own account we see He can do by that other expedient, viz., Christ instead of our perdition.

§ 6. *The Opposite, Fetich.*

This appears in two particulars,—First, that God is made a figment in respect to wrath, and second, by a necessary repercussion of the thought, that He is made a figment in respect to mercy. In neither case is He a genuine Deity. Let us start in Dr. Hodge's own language:—

"If moral perfection be not in Him what it is in us, then He is to us AN UNKNOWN SOMETHING."

We will not repeat, that man has no right to take vengeance. Though Dr. Hodge tells the anecdote of the English judge, yet, after all, men would have no right to hang or to fine but for some sake of defending the community from evil. We will not argue this. We are sure that this much will prevail in the opinions of the people.

But if "*qualis homo talis Deus*," then this is all that belongs to God, and what are we to do with Dr. Hodge? Turrettin, even, does not deal with the stringency that he does. He speaks of justice as *naturalis* (P. i. Lo. iii. Qu: 19). So do we; and that in a most important sense. It is necessary, with our constitution, and has its wisdom in the very *nature* of the case. But what are we to do with vengeance? Notice now its wonderful agreements. We are to have a God all for Himself. We are to have a will of God which is the ground of moral obligation. We are to have an idea of God which is innate, and that idea is a conscious sense of responsibility. Holiness has been left out of the account; and, as we have said, the crust of what is left has been thickened that the idol might stand up. Nothing would be more natural after that than to put on the grim colors. Holiness having been discarded, a trait that would certainly have been sufficient for everything, ministering to the good of the creature, and setting punishment in its healthy place, punishment has to be planted firm at any rate, like horns upon Isis; the idol must have a flattened skull and a distended nostril; and hence the fetich-taint. No one can read these sterner sentences without the fear, that there is something mythic and vengeful like the car of the Hindoo.

And then, mercy! see how that is warped from its natural direction. Beginning everything with the Scriptural idea of holiness; taking its two precepts, and remembering,—"On these two commandments hang all the law and the prophets," it is perfectly easy for us to tell how lost men are saved. God loves them. He has two loves, a love for the welfare of others, and a love for moral excellence. To these two loves there ministers punishment. This it does *ex natura*. Its law is in the essence of things. Its influence demands its use. Its use has been sworn to in a perpetual government. This is the whole story *in nuce*. But who are saved? *We* can say,—All those whom holiness selects; —all those who can bear to be saved consistently with the holiness, and consistently with the happiness, of all the universe. There needs no new affection. Love! That is benevolence. "God so loved the world." That is, wished its welfare. Out of that one throat of benevolence, agreed with, and over-ruled, by a love of holiness, flows all the fountain of eternal peace. We are equipped, therefore. Out of a "moral perfection," which is "in Him what it is in us" (i. p. 420), we have a warm Deity, whom we can look up to and understand as incarnate in the Redeemer. But how if holiness glides out of view?

Dr. Hodge has a problem to settle. Given a hard Christ, who works to display His glory; given a hard right, which is manufactured by the will of the Almighty; given a vengeful ire, which is to satisfy an appetite of God,—how does He draw distinctions? how can He love one and feel a vengeance for another? *We* can settle that easily. He is benevolent towards all. To-morrow would open Tophet and let the Arch Fiend be a trophy of His love but for one grand consentaneous appetite. That appetite is for holiness. Show God how He can save consistently with the happiness and consistently with the holiness of the widest universe, and He will save all; for "His tender mercies are over all His works."

But Dr. Hodge has cut himself off from this. He needs other attributes. Brahm, needing a base for one foot of his crocodile, must fain supply it. And Doctrinalism makes no account of reason. It fables like a Hindoo. God bereft of holiness has to have some reason for destroying one man and saving another. Dr. Hodge supplies it. He speaks of a "peculiar, mysterious, sovereign, immeasurable love, which passes knowledge" (i. p. 549). How that sounds like Vishnoo! He speaks of it as "infinite" (*ib.*). How can that be when it can have five objects instead of one? He says, "it is discriminating, fixed on some and not on others" (*ib.*). And then to unfold his use for it, he says,—"It is to this love, not to general goodness or to mere philanthropy, but to this peculiar and infinite love, the gift of Christ is uniformly referred" (*ib.*). To meet the cavil that such a love must be a fetich, because it clashes with the holiest attribute of character, he boldly grapples with that principle of antagonism, and meets it simply by affirming it. He says,—"It cannot be explained away into mere general benevolence or philanthrophy. This peculiar love of God is not founded upon the fact that its objects are believers, for He loved them as enemies, as ungodly. This representation is so predominant in the Scriptures, namely, that the peculiar love of God to His people, to His Church, to the elect, is the source of the gift of Christ, of the mission of the Holy Spirit, and of all other saving blessings, that it cannot be ignored in any view of the plan and purpose of salvation. With this representation every other statement of the Scriptures must be consistent; and, therefore, the theory that denies this great and precious truth, and which assumes that the love which secured the gift of God's eternal Son was mere benevolence which had all men for its object, many of whom are allowed to perish, is unscriptural" (i. pp. 550, 551).

Now let me be distinctly understood. I do not deny election, and I do not deny electing love; just as I do not deny destroying vengeance. In the language of the East, words are employed to express much under a single expression. Solomon says,— "He that spares the rod hates his son" (Prov. xiii. 24), by which I do not understand that he may not dearly love him. Again, Wisdom says,—"All they that hate me, love death" (Prov. viii. 36); by which she does not mean that they do not hate it.

God has but two moralities,—He loves our welfare, and He loves our holiness. Doubtless He loves the picture of what we will be in His eternal presence. But that He loves us with an infinite affection at the same time that He hates us as unspeakably depraved; that the bandit who has cut my throat, and who at the same time is to be converted on the gallows, is the object of a "peculiar, mysterious, sovereign, immeasurable love" of God (i. p. 549), different from "benevolence" (p. 551), and different from esteem, even from that esteem which is yet to be for him in all the endless ages, is a mere painting of the brain; it is a mere jingle of speech,— a river going up stream and going down stream at a point of time; it is a *nichts-werth*, beyond all manner of doubt; and the Church, bereft of true thoughts of God, in a long decline, has taken this dogma in their place, and hung it like beads upon her Deity.

Because, what can we not find in simple morals? God is pitiful. That covers everybody. He says so. "I say unto you, love your enemies, that ye may be the children of your Father; for He maketh His sun to rise on the evil and on the good and sendeth rain on the just and on the unjust" (Matt. v.

44, 45). Now unite with this, love of holiness; make that His extreme trait, which would sacrifice everything beside,—and what do we want of a "peculiar, mysterious, immeasurable love?" On the principle of parcimony we ought to reject the rest. If God pities all men, that is the *vis a tergo*. If He delights in holiness, that will keep him rigid. Indeed benevolence and this combined will both appeal to Him to be distinctive in His gifts, and to elect the objects of them, without the need of a separate desire.

CHAPTER V.

GOD'S HIGHEST END TO DISPLAY HIS GLORY.

§ 1. *Dr. Hodge's Statement of the Doctrine.*

VOL. I. p. 536. "As God is infinite, and all creatures are as nothing in comparison with Him, it is plain that the revelation of His nature and perfections must be the highest conceivable end of all things." "Whatever He does or permits to be done is done or permitted for the more perfect revelation of His nature and perfections." "The end of creation, therefore, is not merely the glory of God, but the special manifestation of that glory in the person and work of Christ." "Having this great end in view, the revelation of Himself in the person and work of His Son, He purposed to create, to permit the fall, to elect some to be the subjects of His grace and to leave others in their sin" (vol. ii. p. 321).

§ 2. *Contradictions.*

We expounded these under another head (p. 89). God is *right*, Dr. Hodge labors to show, in making His own display His highest end, and thereby seems to admit that *rightness* must be higher than the display. Moreover, he makes God's highest end to be Himself. This, as we have already expounded (p. 89), is a different end from the display; and even if the latter were His delight, it would still have to rest upon the ulterior aim. We must not repeat, however.

§ 3. *Scriptures.*

Dr. Hodge can find abundant Scripture for making God's end His highest "glory." This is the vice of his reasoning. God's "glory" means His *weight*, His *excellence;* for *weight* is the very sense of the word in the ancient Hebrew. God does every thing for His *weight*. That is our very doctrine. The languages of men all trace to matter. *Straightness*,—that grew to mean righteousness. *Levelness*,—that our English oddly renders uprightness, which is a different idea. *Wholeness*,—that we use still, viz., integrity. *Weight* was a capital word. It was a word for merchants. When Isaiah says,—"The king of Assyria and all his glory," it should be,—"all his masses" (Is. viii. 7). The verb meaning, to *be heavy*, and in the Hiphil, to *make heavy*, is the common Hebrew for that thought all through the Old Testament inspirations.

Glory, therefore, is weight. God's doing everything for His own glory means that He does everything for His own excellence. And excellence being simply holiness, we are back at our principle, that the great end of God is His own highest holiness.

This agrees, too, with the end of man. The chief end of man, when it comes to a matter of desire, is God's holiness. But as that is a fixed quantity, and we cannot increase it, our chief practical end is the holiness of the universe. The chief end is holiness. And whether by God or man the great embodiment of that trait is seen to be in the Most High.

If Dr. Hodge declares that there are texts which speak of the *display* as a great end, that we admit. We are treating of the *chief* end. If he says there are advantages in the display of God, that we admit too. They are advantages of the highest kind. Christ teaches us to pray,—" Hallowed be Thy name." But happiness is a great end; and pardon is a great end; and beauty is a great end. We are not commenting on the *great* ends of Heaven; we are only asking whether the *chief* end of God is the display of His own glory.

§ 4. *Argument from Reason.*

And we say it is not,—first, because it makes the chief end of God terminate upon the creature. That is utterly irreverent. Where was the grand chief end in a past eternity? Dr. Hodge argues that God is His chief end, and then He makes His chief end take in a drama before the creature.

Second,—It moves the question, What is God's end in this? Of course it is unphilosophical. A chief end is a chief end. A chief end *of* a chief end is not to be thought of. And yet the proposition,—God's highest end is the display of His own glory, allows all manner of inquiries to come in behind, because we instinctively see that there must be a higher end for showing Himself to the creation. That higher end is His holiness.

§ 5. *Everything else Fetich.*

In a simple worship, God's excellence of character takes the highest place. The humblest adore it. A little child would say, God's end is to do right. And if we address a child and tell him,— God is perfect in Himself: He does not need any of us: He would not have conceived a creation if it was to benefit Him and to build Him up; the little child can follow us through all these ten points of Dr. Hodge. It will fill him with rapture to think that God does everything in order to give way to infinite holiness. But if I tell him, No; He works for Himself; He makes things right just as He prefers; all we are bound to think of is our responsibility to Him; He punishes to satisfy His nature; and we come to be punished and are finally lost to manifest His glory,—his face covers with a cloud, and I talk to him with the jargon of a priest, and put to him conundrums like a heathen worshipper.

CHAPTER VI.

THIS UNIVERSE NOT THE BEST POSSIBLE.

§ 1. *Dr. Hodge States the Doctrine.*

ON page 420 (vol. i.) he condemns the doctrine "that this world, the work of a God of infinite benevolence, wisdom and power, must be the best possible world for the production of happiness." By "*world*" we understand the writer to mean, *universe;* otherwise there is no soundness or fulness to his doctrinal intimation. Hell may not be the best possible for happiness, but the universe it belongs to may be. On pages 432-3 he uses "the world," "the universe" and "the creation," interchangeably. Unless he means that the universe is not the best possible, he means nothing; for no class that we know of, make the world the best possible, separated from what remains. Accordingly Dr. Hodge interchanges the expression in the same sentence ;—
"The universe being the work of God [this is a thought that he is decrying] must be designed and adapted to secure that end (the greatest possible happiness), and is, therefore, the best possible world, or system of things" (ii. p. 145). On page 436 (vol. i.) he extends this denial to holiness. He says,—" We are not obliged to assume that this is the best possible world for the production of happiness, or even for securing THE GREATEST DEGREE OF HOLINESS among rational creatures." See how the Fetich crops out, the grim Juggernaut, not sim-

ply making all things for Himself, but, in doing that, not even producing the best results; erecting Himself into a *se ipso* sovereignty; carving character out of that; breeding our idea of Him as that; and then, when we might expect that the display of His glory would produce the best results, declaring that this rind of a God does not secure even "the greatest degree of holiness" in His rational creation.

§ 2. *Contradictions.*

And yet, if we will watch Dr. Hodge, he cannot persevere in so unreasonable an extreme.

"That God, in revealing Himself, does promote the highest good of His creatures, consistent with the promotion of His own glory, may be admitted" (vol. i. p. 436). Here he advances so far as to concede the beltistic theory except under the caveat that it must not interfere with "His own glory." But in vol. ii. (p. 435) he gets over that point;— "The self-manifestation of God, being the highest possible good," etc. Now this might mean, "good" in itself, just as we have pronounced holiness to be. But he goes further,—"The knowledge of God is eternal life. IT IS FOR CREATURES THE HIGHEST GOOD."

This is the strange thing in Dr. Hodge! that the man chief in dialectic should so double upon his line.

And see further:—"The glory of God is the highest possible end. The knowledge of God is eternal life. It is the source of all holiness and all blessedness to rational creatures" (i. p. 567). "The consequences of the attainment of that end" (viz., "the glory of God") "are undoubtedly the highest good, not necessarily the greatest amount of happiness" (*and he has expressly said,—" not the greatest degree of holiness!"* i. p. 436,) "and that highest good may include much sin and much misery so far as individuals are concerned" (i. p. 568).

§ 3. *Scriptures.*

And all this becomes more palpable when we see what Scriptures are called in. Dr. Hodge rests everything upon the Scriptures that make God's glory the great end of all He does (see vol. i. p. 567). He has a similar argument under the head of Vindicatory Justice. It amounts about to this,—' A red cow is in the paddock near the house, therefore a black horse is not in that paddock:' or, ' I saw him eat peas for dinner, and therefore he certainly did not take mustard.' The argument in case of Justice is,—"The amputation of a crushed limb is not of the nature of punishment," therefore punishment is not an instrument, "and its special design is not the good of society" (vol. ii. p. 578). That is,—'A hoe is not for the good of the garden, because a hoe is for the killing of weeds. And the reason will immediately appear when we remember that a rake is for the good of the garden, and is never the least for the killing of weeds.'

So in this instance of the universe. "The only satisfactory method of determining the question is by appealing to the Scriptures" (i. p. 567). Well, what do the Scriptures teach? "The Scriptures teach that the glory of God is the end to which all other ends are subordinate" (i. p. 435). Agreed; and what do we learn from that? Why, that "we are not obliged to assume that this is the best possible world for the production of happiness, or even for securing the greatest degree of holiness among rational creatures" (i. p. 436). A greater *non sequitur* we never dreamed of. And yet this is the only argument. These are the only Scriptures that Dr. Hodge brings upon his page. "Of Him and through Him and to Him are all things;" and, therefore, this is not the best possible universe. This is the gist of the proof that Dr. Hodge brings out of God's Holy Word.

§ 4. *Reason.*

But out of *Reason* the case is somewhat different. Here Dr. Hodge has two arguments,—one built upon the finite, the other built upon the prevalence of sin. He charges the optimist with blasphemy. Here, says he, is a finite universe. It is little. It might be uttered at a breath. To say that God is limited, and is at the end of His resource, is to say that He is not Omnipotent. And again, to say that He is bound down so that He must submit to a universe with sin, is to say that He could not create it differently. Any May morning He could roll forth forty worlds freighted with higher happiness, and by the word of His power He could have kept iniquity out of His sight, and banished it forever from His creation (see vol. i. p. 433 *et alibi*).

This seems unanswerable.

But we can frame the very opposite argument. Suppose God should *want* to make a beltistic universe: suppose He longed for it: suppose He bent towards it with all the ardors of the infinite: according to Dr. Hodge He could not do it. Does that not limit His omnipotence?

God *is* limited.

That, perhaps, is the boldest way to answer Dr. Hodge.

He is limited forever and forever; and yet He is still omnipotent. And if you ask me how, I say,—He is limited by the impossible. He cannot make man God. He cannot make the universe Divine.

Dr. Hodge says, He can roll forth forty worlds: but he can say that to-morrow and to-morrow. He can stand on the verge of space, and when the universe is as near infinite as he can possibly conceive, he can cry out—Forty more, and, Forty more. There is no limit to this appeal.

But there is a limit to the possibility.

And here precisely is our argument. The universe is of its very nature finite. This implies an end. This implies that God must set it. Suppose He wished it the best possible. It would certainly deny His Omnipotence if He could not have it. And yet, if He had it, it must be still finite. God could ordain the boundary; and one boundary could be wise, and another foolish, and this is all we declare. God has made the best possible universe; and by this we mean the holiest possible and the happiest possible for the vast creation.

And as to the other argument, how singularly Dr. Hodge contradicts himself. He says, so teaching we deny God's omnipotence. God any morning might create a world without the incubus of sin. Who would ever dream that the world gains by sin, and that this is a teaching of Dr. Hodge? But we have only to listen. " As sentient creatures are necessary for the manifestation of God's benevolence, so there could be no manifestation of His mercy without misery, or of His grace and justice if there were no sin. The knowledge of God is eternal life. It is for creatures the highest good. Sin, therefore, according to the Scriptures, is permitted, that the justice of God may be known in its punishment, and His grace in its forgiveness. And the universe without the knowledge of these attributes, would be like the earth without the light of the sun " (i. p. 435). If the universe is the better for sin, that does not make it the best; but if the universe is the better, that makes sin to be no bar, and indeed takes it as a help, and makes it lift that much nearer our required consummation.

And yet Dr. Hodge is so inconsistent; for though he admits this, yet listen to him again:—" Sin, in itself, is an evil ; relatively it is a good." (Dr. Hodge is ridiculing the doctrine of those, who, nevertheless, in moral consequences seem just where he is.) " The universe is better with it than without it. In itself it is an evil that the smaller animals should be devoured by the larger; but as this is necessary to prevent the undue development of animal life, and as it ministers to the higher forms thereof, it becomes a benevolent arrangement. The amputation of a limb is an evil; but if necessary to save life, it is a good. Wars etc., ... Thus if sin be the necessary means of the greatest good, it ceases to be an evil on the whole, and it is perfectly consistent with the benevolence of God to permit its occurrence" (i. pp. 432, 433),—a conclusion here at the very climax of his irony which seems in just intent the same with his own as previously quoted.

§ 5. *An Orthodox Optimism.*

Holiness is the only thing that can bring us safe all through this labyrinth. Holiness is a love of others and a love of the quality of holiness itself. Of these the latter is the more imperial. Holiness, therefore, is the great object of God in all His creation. He loves His own holiness better than that of any of the creatures. But unless we can suppose that His own holiness, which is a love of others and a love of the principle of moral excellence, can interfere with the holiness of other beings, we may easily infer that *quoad* the creature God's highest object is the holiness of all. Now if God's highest object *ad extrâ* be the holiness of all, is it, or is it not, gratified? Or if for wise reasons He prefers the highest holiness of the most to a lower holiness of all, has He or has He not His wish? The demonstration seems complete.

But if the holiest possible universe results, could that be a less happy one? The inference seems plain, that if God be a holy God the universe that He has made is the holiest, and on that very account the happiest, that could possibly come into being.

§ 6. *The Opposite, Fetich.*

We have hinted at this already (see p. 26).

But a Deity erected over the universe without a life to the benefit of the universe, is a grim idol. There is an unnaturalness in Him in the very thought. A mere shell-like God!—a Deity for the mere sake of a Deity! a sort of *semet ipso* Jehovah! why, the world is full of such hybrids. The Pantheon was crowded with them. We believe that Doctrine can beget a monster as vile as Polytheism. For see the different traits:—*Power;*—that breathed from Olympus; *Wrath;*—that burned in Moloch; *Self-Providence;*—that pampered itself in Mammon. The God not weaving out the happiest results is the Ormuzd beset by His Ahriman. All wears an idolatrous phase. All breaks into detail and refuses unity. The only One God is the God infinitely holy; who, therefore, makes holiness the object in all His works; who, therefore, makes the sum of creatures to the uttermost holy; who, in so doing, makes them to the uttermost happy; and who, when He has once given them a perception of this, fills their lips with an adoration that it is hard to think of as being ever curtailed.

CHAPTER VII.

GOD'S PROVIDENCE NOT A CONTINUOUS CREATION, ELSE GOD THE AUTHOR OF SIN.

§ 1. *Dr. Hodge's Statement of the Doctrine.*

THE error, as Dr. Hodge regards it, of the theory of a "continuous creation" he gives in three forms. We have not room for all of them. We give the worst. If he is mistaken in regard to that, *a fortiori* must he be in respect of the other two, one of which he refers to the Reformed theologians, and quotes as holding it Heidegger, Ryssenius and, virtually, Turrettin (vol. i. p. 577). Dr. Hodge himself shall describe the form we shall allude to :—" By continued creation is meant that all efficiency is in God; that all effects are to be referred to His agency. As there was no co-operation in calling the world out of nothing, so there is no co-operation of second causes in its continuance and operations. God creates, as it were, *de novo* at each instant the universe, as at that moment it actually is" (i. p. 578). President Edwards has a kindred theory. Dr. Hodge quotes him as saying that "the existence of created substance in each successive moment [is] wholly the effect of God's immediate power in that moment, without any dependence on prior existence, as much as the first creation out of nothing" (vol. ii. p. 217).

Dr. Hodge argues against all this, that it destroys, first, "All continuity of existence;" second, "all evidence of the existence of an external world;"

third, "second causes;" that, fourth, it is like Pantheism; and, fifth, that it makes God the author of sin.

It is very clear that that fifth difficulty is the awakening one with Dr. Hodge.

This appears from his self-contradictions:—

§ 2. *Dr. Hodge's Contradictions.*

He is not sufficiently impressed with the philosophical objections in the case to rest under their consistent influence. Dr. Hodge is ruled by his theology. Imputation coming afterwards to be advanced, and to be advanced in a rigid form of it, he comes himself to need a form of divine creationism, and he seizes it without a scruple. For example, a bean-stalk growing in the night he makes the direct work of the Almighty. A stalagmite growing in a cave; that is nature. God only supports it. The shooting of a crystal or the drawing of a magnet: that all springs from natural cause. To say that God creates all this each moment would be Pantheism. But to say, that He creates life each moment, and that all that breathes lives by His immediate efficacy, there he finds no check. He catches sight of a bearing, not only upon the creation of our spirits, but upon certain infidel schemes, and it warps his thought. Life is God's work each moment. "Life is not the product of physical causes. We know not that its origin is in any case due to any cause other than the immediate power of God" (vol. ii. p. 74; see also vol. i. p. 612). That is, a crystal is not created each moment, because that destroys second causes (i. p. 579); but life is created each moment, and by an immediate "divine efficiency" (ii. p. 74), the only difference that we can see being, that the first is needed to keep God from being the Author of sin, and the last to keep the sinner from being traduced from Adam (see i. p. 70).

A stranger discrepance occurs as between two other tenets. "*Continuous creation*" in one form of it calls for the remembrance of the fact that God is eternal, and that His acts, therefore, are not successive. Dr. Hodge very properly challenges this. "As to the idea that God's acts are not successive; that He never does in time what He does not do from eternity, it is obvious that such language has for us no meaning. We know that God acts; that He does produce successive effects; and that so far as we are concerned, and so far as the representations of Scripture are concerned, our relation to God and the relation of the world to Him, are precisely what they would be if His acts were really successive" (i. p. 578). And yet Edwards' theory,—a much better theory,—comes to be discussed, and Dr. Hodge uses this language:—"It proceeds upon the assumption that we can understand the relation of the efficiency of God to the effects produced in time. Because every new effect which we produce is due to a new exercise of our efficiency, it is assumed that such must be the case with God. He, however, inhabits eternity. With Him there is no distinction between the past and future. All things are equally present to Him. It is surely just as conceivable or intelligible that God should will the continuous existence of the things which He creates, as that He should create them anew at every successive moment" (ii. p. 219).

Our misgiving is, that Dr. Hodge charges himself more about the argument as to sin than as to any philosophic difficulties. Perhaps that is praiseworthy. But it only makes it more imperative that this delicate defence of God should not on that very account, and because of its look of partiality, in its effect impeach and betray Him (*see Commentary on Proverbs*, pp. 401, 402). Because—

§ 3. *Argument from Reason.*

Dr. Hodge has involved beforehand the character of God. He has said,—Unless a certain philosophical theory can be maintained, God is the Author of sin. That theory is, that God directs and sustains but does not continuously create His finite creatures. His philosophical arguments are of necessity false: I mean by that, one theory claims all the facts like the other. When Dr. Hodge says, that one theory destroys identity, or, as he expresses it, "continued existence," he only means the truism, that identity under the one theory is not the same as identity under the other. So of " second causes." The argument is nothing. Both theories use the same language. Both theories proclaim the same facts. A cause when existing in the creature is certainly different from a cause found in the Creator, and has but the faintest resemblance in answering to the word; and yet where does Dr. Hodge notice that fact? Being! What is being? Being, when it is Divine, and being, when it is the work of the Almighty! does *there* appear no difference? And, therefore, what is all this reasoning worth? The being of God is different from the being of man; now, who shall say, how different? They might be called by different names. Being is but a word in the dark; and all men must say that it is real chiefly in the Almighty. And now, what our being means, and what our effecting anything means, are precisely what these theories we are discussing shall enable us to determine.

How idle to declare that either theory destroys this or that. Either theory destroys the other. Neither theory destroys identity, or continued existence, or secondary causes. It only affects their shape. As long as a continued-creationist speaks of personal identity, and compares it to light upon a portrait (see Edwards, vol. ii. p. 555); as long as he maintains causation, and expounds it; as long as he insists upon existence, but only makes it different from God's,—how futile to say that he destroys them, particularly as all alike declare that in these things we are very different from the Almighty.

Dr. Hodge is thrown simply upon his ethical conceit.

When in old temperance days leading divines declared that if Jesus of Nazareth created intoxicating wine he was a brutal impostor, all temperate thinkers stood aghast. Ought we not to risk less in our theological assaults? Dr. Hodge has let slip the arrow that God is the Author of sin. It is true that he is aiming at some other God, but men of equal prayer think that he is aiming directly at their Deity. Was it distinctly prudent to shoot it? Now that the philosophic reasons seem of little force, let us see to what narrow ground he is confined for his vindication.

He teaches (vol. ii. p. 70) that the soul of an infant is immediately created by the Almighty. The body which is derived from Adam can have no fault, and cannot carry blame or wickedness. What is the result, therefore? The result is that God creates a poor infant wicked. Why does He do it? Because Adam was wicked. When did Adam live? Six thousand years ago. Why does He make the infant wicked? Because of a federal covenant. Is God unjust in this? No: it is impressed in the creation; all nature wears its analogies. Dr. Hodge in expounding this tells only the truth, and all orthodox men will look upon it with pleasure.

But now see the difference. God creates *every* soul directly. He only hath immortality. He gives being every moment. He not only creates the child, and curses him for Adam, but He creates the man, and curses him for himself; where will you distinguish? Dr. Hodge not only does not think that God is the Author of sin because He creates what is wicked, but he does not think it even hard that He creates a lost thing *de novo* for a trespass six thousand years ago. And yet if I teach that there is a continuous creation, and like the light perpetually on the moon there is an immanence of God's power that keeps the child, and gives him a continual creation; if I say, he cannot have self-being, but he can have being, and it can be kept in him *de novo* and all the time, Dr. Hodge, who has been tolerant of the federal curse, closes upon me at once; utters certain philosophic challenges which we might easily bear, but tramples upon me most for my impugnment of the Almighty.

Now is this prudent?

Take any plain man and say, Here is a child flashed into being *de novo* and with no paternity for his spirit; here is a man flashed into being continuously as the only way in which he can have an existence; the child is created wicked, and the man is created wicked; the child for sin on the part of Adam, and the man for his own sin, that is, for distinctly remembered iniquity;—now our auditor might smile at both, and ridicule one point and another;—but suppose I were to say, there is a writer who denounces this last, and says it makes God the Author of sin, but he defends the former, and says it is all just and right,—how would the peasant man break out!

And it is unfortunate all through; because, first, it settles things by philosophy!

Hear Dr. Hodge himself!—

"It is obviously most unreasonable and presumptuous, as well as dangerous, to make a theory as to the origin" [suppose we say, *continuance*] "of the soul the ground of a doctrine so fundamental to the Christian system as that of original" [suppose we say, *the authorship of*] "sin. Yet we see theologians, ancient and modern, boldly asserting that if their doctrine of derivation, and the consequent numerical sameness of substance in all men, be not admitted, then original sin is impossible." "This is done even by those who protest against introducing philosophy into theology, utterly unconscious, as it would seem, that they themselves occupy, *quoad hoc*, the same ground with the rationalists. They will not believe in hereditary depravity unless they can explain the mode of its transmission." [Let me alter that,—'They will not believe in responsibility for sin unless they can explain the continuance of being.'] "There can be no such thing, they say, as hereditary depravity unless the soul of the child is the same numerical substance as the soul of the parent." [*Alias*—'God Himself must be responsible for sin or else the soul of the sinner must have my special theory for its continuance in being.'] "No man has a right to hang the millstone of his philosophy around the neck of the truth of God" (vol. ii. p. 73). So much for our first point.

But, secondly, it is not only a *philosophy*, but one of a most *precarious kind*. See what it has to establish. God may ordain, and then He is all right. He may create. He may have a universal system of Providence. He may carry this so far that He must uphold and appear and concur in all existences.

Job cries, "Look away from him that he may cease" (Job xiv. 6). All this is agreed in. God must flow-in with His power into every creation, and do it with unslumbering constancy, or that creature vanishes. Dr. Hodge is all clear that far, and there is not a breath upon the wave. But now, one hand-breadth further:—God must continuously create. As all life according to Dr. Hodge is by His direct efficiency (ii. p. 74), so all being is. He cannot relegate it. It must keep flowing. It must be immanent forever. We say this, and we are gone! This that has scarce one point of advance; that has no true moral difference; which can hardly be distinguished by Dr. Hodge if he keeps cause separate from Cause, and being separate from the Great Being; this, which seems consistent with Bible speech, for He forms the light and creates darkness; He makes peace and creates evil (Is. xlv. 7); "By Him all things consist" (Col. i. 17); We live in Him and move in Him and have our Being (Acts xvii. 28); He has "power over the clay to make one vessel unto honor and another unto dishonor" (Rom. ix. 21); He hath made "even the wicked for the day of evil" (Prov. xvi. 4); this theory, which, to say the very most, gives God authority and control and power and foreordination no particle more than the other, is to be its practical antipodes as to making God responsible for our iniquity.

No plain man can see it.

And now we come to a graver idea:—

§ 4. *The Doctrine, Fetich.*

Mark one fact about Dr. Hodge,—He is making a refuge for God behind brute substance.

His apology for God is not only (1) a philosophical one, and not only (2) a tenuous philosophical one, but it is (3) a resort, in a purely ethical question, to a sort of *tertium brutum* of existence. No one denies that matter exists. No one supposes but that God creates it. No one is making a difficulty of His entirely controlling it. And, therefore, when God is all open in every form to an entire association with His works, Dr. Hodge would attempt to defend Him by the veil of some separate substance.

We beg a glance at this.

Men have worshipped substance when they have ascribed to it Divine traits. But men have worshipped substance equally, and perhaps more fatally, when they have ascribed to it infamous traits, and professed thereby to defend the Deity. The Persians did all this when they invented Ahriman as an excuse for Ormuzd. So did the Platonists. Those old theories by which matter was made to be eternal, and became the scape-goat as the origin of evil, were but the instances of the abandonment of God. The Deity does not ask such defence. It has the effect of an excuse for wickedness. Men see the hollowness of it. God all for Himself tempts men because that is not what they admire. And God argued for in these ways is a terrible snare to them, because they see that there is no difference (I mean *in foro morale*) between a God so sustaining that a thing would vanish if He looked away (Job xiv. 6), and a God continually at work, to pour-in the being which He has started in His creature.

If that is an idol which the Hindoos began with as God's work and ended with as—God, that is an idol too, which Dr. Hodge realizes as sustained by the Almighty, but uses as His defence at last to excuse Him from being responsible for evil.

CHAPTER VIII.

MAN'S HELPLESSNESS NOT DISINCLINATION.

§ 1. *Dr. Hodge's Statement of the Doctrine.*

"THE inability of sinners is not mere disinclination or aversion to what is good. This disinclination exists, but it is not the ultimate fact. There must be some cause or reason for it. As God and Christ are infinitely lovely, the fact that sinners do not love them, is not accounted for by saying that they are not inclined to delight in infinite excellence. That is only stating the same thing in different words. If a man does not perceive the beauty of a work of art, or of a literary production, it is no solution of the fact to say that he has no inclination for such forms of beauty. Why is it that what is beautiful in itself, and in the judgment of all competent judges, is without form or comeliness in his eyes? Why is it that the supreme excellence of God, and all that makes Christ the chief among ten thousand and the one altogether lovely in the sight of men and angels, awaken no corresponding feelings in the unregenerate heart? The inability of the sinner, therefore, neither consists in his disinclination to good, nor does it arise exclusively from that source" (ii. p. 261).

§ 2. *Dr. Hodge's Immediate Contradictions.*

I. The seeds of self-rectification are found in these very sentences themselves. Our position is that helplessness is disinclination. Dr. Hodge's is that helplessness is not disinclination. If I could induce an arguer to say that my position was not true and rest there, I should have gained a great victory. What I dread is that he should go on and prove it. But if after saying that it was not true, he should lose his mind for a moment and say it was a truism, I should more than conquer. My forces would have gained the field without the peril of a victory.

"This disinclination exists," says Dr. Hodge, "but it is not the ultimate fact" (ii. p. 261); and yet that it exists and *is* the ultimate fact is the very theory that is asking for refutation. Dr. Hodge brings the opposite in view as though it shone by its own light. But before we can set it up, and plead against it, he bowls it down himself, for he remorselessly declares that what opposes it is a naked truism.

Let us not confuse ourselves by this queer beginning.

Either or either not helplessness is disinclination. If it is, how stands the plea that it is not the ultimate fact? and if it is not, how odd the argument that it is a naked truism!

Let us beg a thorough examination. "Disinclination is not the ultimate fact" (p. 261). This speech stands unsupported. Again, "There must be some cause or reason for it." "Reason" of a historic kind, or "cause" or history why man came into his fallen state, that obviously we are not considering. The "cause" intended must be a psychologic cause. The "reason" must be like faith as Dr. Hodge defines it (iii. p. 41), as the "cause" or "reason" of spiritual repentance. It is one state gendering, or being origin to, the existence of another. And as we mean to deny that helplessness is the occasion of disinclination, to say that the latter is "not the ultimate fact" (ii. p. 261) is the sheerest form of a *petitio principii*:—because, there is the very question. But when we are led farther, and told that the fact that it is is a naked truism, we stare with astonishment. Two and two are not four, and, moreover, the assertion that they are is all a truism. We can stand such attacks. Do not think we are misrepresenting anything, for observe the language (p. 261):—"Disinclination exists, but it is not the ultimate fact. . . . The fact that sinners do not love God and Christ is not accounted for by saying that they are not inclined to delight in infinite excellence. That is only stating the same thing in different words. . . . . The proposition that a man can love God if he will . . . . if the word [will] be taken in a wide sense as including the affections, is a truism" (ii. p. 266). This feels like milder rhetoric than O'Connell heaped upon the fish-woman; for though he called her a "hypotheneuse," and a "mere parallelogram," yet, as far as could be seen, these epithets could neither hurt nor help; but truisms are comfortable helps, and in these earlier stages at least, are a good thing to put in one's first foundation.

II. A second mistake of Dr. Hodge is a sudden transference of inclination into volition. Who agreed to that? I would not like after that to attempt any debate. If I say, I am helpless to eat apples, and then go on to explain that my helplessness consists in disinclination, the man who holds that this is "not the ultimate fact" because I cannot will to change my appetite, and therefore there must be something back of disinclination, is changing under the very eye of day inclination into will.

Dr. Hodge does it.

"If the word *will* be here taken in its ordinary sense for the power of self-determination, the proposition that a man can love God if he will is not true, for it is notorious that the affections are not under the power of the will. If the word be taken in a wide sense as including the affections, the proposition is a truism" (p. 266).

No one ever held that a man can change hate into love by an act of the will.

But though Dr. Hodge cannot prove that helplessness is not disinclination by showing that a man cannot will to love God, we can prove that it is disinclination by showing certain bearings of will of a secondary and instrumental kind.

What is will?

All will is inclination; but all inclination is not specifically will.

I cannot will to love God. Why? Because the will does not do such things. It moves the muscles, and it is concerned in the act of attention. The only way I can love is to love. It is as Dr. Hodge would deny, an ultimate fact. To will to love would be about like digesting tunes, or listening to a poison. Yet though unwarrantably brought into the debate when no man fathers it, and no notice has been given of its taking the place of inclination, yet now that it has been brought in, we have a use for it, and a very welcome light which we derive from that very confusion that has sprung up in Dr. Hodge's mind.

We cannot will to love God, and our only way to love Him is to love Him, and our helplessness to love Him is that very want of love, or as Dr. Hodge denounces it, the very fact of our disinclination.

But though we cannot will to love God, we may, for shortness of speech, speak of that very thing.

There are two ways of accomplishing His love, one to do it at once, which we cannot unless we love Him; the other to USE THE MEANS.

Now that man's helplessness consists in disinclination appears, most patently of all, in this, that he cannot even use the means. He can if he will. But there is the gnomon to the whole. I cannot love God without the use of means. Why? Simply because I do not love Him, or in other language, because I am disinclined. Again, I cannot love God through the use of means. Why? Simply because I will not use them. My helplessness is total. My love itself, that is a thing utterly and profoundly dead; and my steps to get it, they are on that very account *nil*. All will is inclination, and if my inclination is entirely the other way, I am utterly unable to use the means to secure a love for my Creator.

III. Now, strangest of all, Dr. Hodge is not orthodox when it comes to this. Building his thought on the idea that I can do certain things if I will; remembering the stringency of his speech, that man's helplessness is not disinclination; cut off from our explanation of the fact that I cannot use even the means of grace,—he shocks us suddenly by abandoning that asseveration.

The sinner "can be kind and just . . . . and [has an] inability [that can be] asserted [only of] the things of the Spirit" (vol. ii. p. 263). Again, in respect to the use of means, read all on pages 276 and 277. "When a man is convinced that the attainment of a desirable end is beyond the compass of his own powers, he instinctively seeks help out of himself. If ill . . . he sends for a physician. If persuaded that the disease is entirely under his own control, and, especially, if any metaphysician could persuade him that all illness is an idea, which can be banished by a volition" [I know of no mortal of whom that illustrates the creed], "then it would be folly in him to seek aid from abroad. The blind, the deaf, the leprous and the maimed, who were on earth when Christ was present in the flesh, knew that they could not heal themselves, and, therefore, they went to Him for help" (p. 277).

Now we believe that they could not go "to Him for help."

Our notion is of a *total helplessness.*

We believe that they cannot stand in the tower; and that far the reason is simply that they do not stand there. But we believe further that they cannot climb, even though there be a stair, and cannot shout, even though there be help; nay, that they cannot even look that way. They can as possibly leap into the very top as lift a finger to get there; but all because of a disinclination.

See, like an old machine! a pivot has cut too far one way,—as, that being helpless is not being disinclined,—and presently it has cut too far the other way, and being helpless is not being helpless in every conceivable particular at all. A machine too tight, works logically into being far too loose. Ours is a perfect helplessness. But we began it in the beginning as the aversation of the soul, and so we hold it to the end as a not thinking well, or a not being inclined, to use the means for its own salvation.

§ 3. *Scripture.*

It proves nothing to show that the soul is entirely helpless. Dr. Hodge's Scripture (vol. ii. p.

268) is all of that class. We hold a more total helplessness than is maintained by Dr. Hodge. That "the branch cannot bear fruit of itself;" that "without [Christ, we] can do nothing" (John xv. 4, 5); that "our sufficiency is of God" (2 Cor. iii. 5); that "no man can come to [Christ] except the Father draw him" (John vi. 44); that we are "not subject to the law of God, neither indeed can be ; so then [that] they that are in the flesh cannot please God" (Rom. viii. 7),—ought to have been quoted by Dr. Hodge first, to show the doctrine of our helplessness, and then, after having made sure of that, it must be a different list that can show that helplessness is not disinclination.

Let us quote some on our side of the question. "Ye will not come to me that ye might have life" (John v. 40). David says, "God shall wound the head only of His enemies" (Ps. lxviii. 21 ; *for* "only" *see Hebrew*); and again, "Only the rebellious dwell in a dry land" (Ps. lxviii. 6). Ezekiel implies that it is all within our own governance,—"Why will ye die?" (Jer. xxvii. 13 ; Ezek. xviii. 31 ; xxxiii. 11). "How often would I have gathered, etc., etc., but ye would not" (Matt. xxiii. 37). "Even as they did not like to retain God in their knowledge" (Rom. i. 28).

Dr. Hodge's great text will introduce us finely to the next section.

§ 4. *Argument from Reason.*

Dr. Hodge's great text is, "The natural man receiveth not the things of the Spirit of God ; for they are foolishness unto him ; neither can he know them because they are spiritually discerned" (1 Cor. ii. 14).

First; as to the text. It describes the very kernel of our helplessness. We are helpless because we are dark. But now let us be careful. What does that demonstrate? It demonstrates, beyond a doubt, that the natural man cannot see; but does it demonstrate—I mean by this text alone—that thereanent, or rather as therein contained, it is not also true that he cannot feel? If I saw a whale in a certain sea, does it forbid that I also saw a porpoise? Or, coming a great deal nearer, if I saw a beauty in a certain picture, is it more or otherwise than that I simply felt a beauty?

Dr. Hodge has switched off his rail-train precisely at this cut across the mountain.

I see first; I feel afterward. That is his imagination.

Then if I see first and then feel, of course I am helpless to feel until I can see.

And then, as my seeing first depends upon the Almighty, my helplessness does not consist in disinclination, but in a want of seeing.

Suspicion gathers at once. Dr. Hodge says, "No man can see the beauty of a work of art without æsthetic discernment" (vol. ii. p. 269). Put that a little differently:—'No one can detect water without aqueous intelligence.' What does that mean? Again, "No man can see the beauty of spiritual things without spiritual discernment" (p. 269).

But, now to go back; we are shown beyond

doubt that the sinner's trouble is a want of vision. Moreover, that is his helplessness. Sinners cannot be saints without seeing, and seeing cannot be had without the influences of God's Holy Spirit. Unquestionably he is right in this, and, one might think, right in his whole argument. But those sentences so truth-resembling above, are the pointers to the unobserved delusion.

Vision is helplessness; but what is vision?

Vision is the gift of God's Holy Spirit. Vision is a new life. It is total grace; and the whole of piety. Give a man a ray and it saves him, for "this is life eternal, that they might know thee the only true God, and Jesus Christ whom thou hath sent" (John xvii. 3).

But before I can go further and say,—A want of vision being our helplessness, and a want of love not being a want of vision, a want of vision must precede a want of love, and the whole region of affection is not the region of our helplessness, we must look first at a *non sequitur* in this proof, which has surreptitiously clambered in.

Who says that want of love is not a want of vision?

This fancy has been a Trojan horse.

"Rather" (says Dr. Jackson, of Corpus Christi College, Oxford), "this erroneous imagination of such a distinction between these faculties" ["of will and understanding"] "hath plunged the maintainers of it in such soul-inextricable errors, and driven them to such miserable endless shifts in matters moral and theological of greater consequence, as every intelligent man should abhor" (Justifying Faith, 1631, p. 49).

So let us look now to a regular argument. We do not deny that being dark is the very essence of being helpless; but we do deny that being dark is not also the essence of being unloving. Having no inclination to love God, and having no illumination to see Him, are not, as concerns the genesis of both, the consequences the one of the other, but are the same individual condition, distinguishable as aspects of the fact, but logically or chronologically having neither precedence.

Try this on different expressions:—I perceive warmth, and I feel warm: are these different numerically? I see the beauty of a poem, and I feel the beauty of a poem, and I love its beautifulness. Are those *in esse rei* numerically different? Now climb up with that a little higher, and try it in ethical affairs. I see the excellence of God, and I feel the excellence of God, and I love the excellence of God, or, if you please, I love God: are those sequent, the one of the other? or are they all a unity? Understand me. I do not mean, synonymous, or that they are all expressive of the like, but that they are all one state, turned in the aspects of our speech and set to a selected phase.

So in the case before us.

Holiness has three senses: either, first, a quality; or, second, the things in which this quality appears; or, third, a character.

In the second of these senses it applies to two emotions: either, first, a love of others' welfare; or, second, a love to holiness in the first sense, viz., a love to the quality that belongs to these holy things themselves.*

Holy things, therefore, are feelings, viz., (1) Benevolence, and (2) Love to the quality of virtue; and correspond in every particular to what Christ marked (Matt. xxii. 40) as the two tables of the law.

But now, if holiness be these two emotions, then sin is the want of them. There is no sin in hell or earth that is not the want of benevolence or the want of love to the quality of right.

But if sinfulness be a want of love, pray what is helplessness? Is helplessness any different from sinfulness? I do not mean, does it mean differently? but is the thing different? Nevertheless, if helplessness be the same as sinfulness, and sinfulness be a want of love, we have our conclusion beyond a doubt, that helplessness consists in disinclination.

We accept, therefore, Dr. Hodge's illustration greedily. A man cannot love a picture till he sees its beauty (ii. p. 269); but will any man point out to me where these two things separate, and become either source or consequence? I do not taste first, and enjoy afterward; but I enjoy in tasting. Loving the warmth of my cloak, and knowing the warmth

---

* These are statements in which we draw a little upon our ethics, but not illogically so, since Christ's language (see Matt. xxii. 40) affords the same premise.

of my cloak, and feeling the warmth of my cloak, are discrepant in speech, but he is a sharp thinker who can look between them.

This is no trifle, this *lumen siccus* antecedent to affection. It appears worse under the category of Faith (see iii. p. 93), but bad, manufactured into helplessness. It builds a motive for delay. Skeptics like it. It makes the offers of the gospel a sort of trifling. Reason scoffs at such assurances of welcome. And when Dr. Hodge proceeds, and separates faith from morals; when he says,—" There is an obvious difference between morals and religion " (vol. iii. p. 63); when he says,—" The difference between these two classes of acts, although it may not be easy to state, is universally recognized;" and when he marks it still closer, as " between those religious affections of reverence and gratitude which all men experience, and true religion" (*ibid*), we are almost in despair.

Let us be careful, therefore.

There is a sense in which such sentences have always been accepted. Morality, in the worldly view, is very impious. But why is this? Because it is a morality that is so little moral. Morality as really moral and yet not identical with holiness; the moral law rigidly observed not the part of the believer; observing it better not the mark of his good estate; one virtue for the street and one virtue for the house of prayer; pure religion in the opinion of God not alms-deeds (Jas. i. 27); and men not promised that their sins shall be as snow if they put away the evil of their doings from before their eyes (Is. i. 16, 18),—is a hand-writing of the age, which seems to show why there are so many culprits in the church, and to erect a Juggernaut of faith in which inability to repent is something else than wilful wickedness.

We believe in inability to repent, if anything more totally than is believed by Dr. Hodge (see vol. ii. pp. 275-277). The sinner will be certain to repent if he perseveres in using the means. Religion is a more certain enterprise than farming if we employ diligently the means of grace. The door is wide open to the worst; but then they must employ with honest earnestness the means that are offered to every one of us. This they will not do. Herein consists their inability. We do not believe that men are helpless to love God but helpful to employ the ordinary means: we believe that a common helplessness lies over every impulse. We believe this is a helplessness of will when will is the thing concerned. To speak of will we must look at some province of volition. The love of God is not such a province; and there our helplessness is sheer disinclination. The use of the means of grace is such a province; and there our helplessness is disinclination also. To say the opposite is fetich. And when Dr. Hodge, after actually quoting the Confession (chap. vi.), "utterly indisposed . . . and made opposite to all good, and wholly inclined to all evil" (see vol. ii. p. 260), goes on to say that we are commanded to repent with the distinct knowledge that we cannot do it (ii. p. 271, 272), and that that inability does not consist in disinclination, and that then our own consciousness recommends this as actually the fact! (p. 273,)—he is griming his own idol, like an Aztec priest, and telling his own story in a way utterly irreconcilable with our real wickedness.

Hell is for not loving right and for not loving the welfare of those around us. When a man does not love a thing, he cannot love it, which Dr. Hodge justly calls an identical proposition ; but then that inability is an inability of the affection, in other words it is disinclination. For those thus disinclined open doors are set of easy rescue. They cannot take them. Why? Simply because they are not inclined. This makes their guiltiness. To say that they have an inability farther back, and that God knows it when He tells us what to do; and that " the command is nothing more than the authoritative declaration of what is obligatory " (p. 267) ; and then that our consciousness justifies all this (pp. 273-5),—is just of that crust of unpracticableness that makes God all for Himself, that gives us an innate idea of His supremacy, that makes His revenge as of His highest taste, that makes Him ruin in order for display, and that makes Him arbiter of right and able to give it its nature by an action of His will.

CHAPTER IX.
SAVING FAITH NOT IN ITS ESSENCE MORAL.

BUT, now, when all this comes into the region of Faith, the idol is grimier yet.

§ 1. *Dr. Hodge's Statement of the Doctrine.*

A man who takes in Dr. Hodge's definition of God, that He is a being that does all things for Himself; and who takes in his genesis of right, that it is a right made right simply by the will of his Creator; and then the genesis of our idea of a Creator, viz., that it is an innate idea, and that the idea is an idea of responsibility and dependence,—will be prepared for his definition of faith, viz., a trust to Christ on this Supreme Person's testimony. The grim feature would hardly be kept up unless a trust to the Supreme on a *semet ipso* authority and will shrouded again His intrinsic excellencies.

We believe that saving faith is a trust to Christ on a sight of His excellency. Dr. Hodge believes that it is a trust in Christ nakedly on testimony.

We believe that holiness is of the very essence of faith. Dr. Hodge believes that faith is a sheer thing by itself, and holiness must be kept out of it and be its sequent.

Narrowing it in this way, we might at least claim that he should be precise. But here is the suspicious circumstance, as it is with all idolatry, that as we approach the Stonehenge circle the gloom of oaks gathers about us. We defy any one to tell exactly what the *differentia* of Dr. Hodge's faith can be succinctly imagined.

Men identify saving faith by two aspects of it, (1) its nature, and (2) its object.

Dr. Hodge tells us nothing that marks the *differentia* of faith, except that it has the testimony of the Spirit. He tells us, the Spirit is invisible, and that the testimony is known only by its effects. (1) The *nature*, therefore, is not revealed by this ; for *belief* and *trust*, and whatsoever other psychological word is employed for the effect of the Spirit, has no *differentia ex se*, for the reason that precisely the same words are used for other exercises.

We give a mark boldly. We say the *differentia* of saving faith is holiness. All faith before that is common. Upon mounting to saving faith it becomes holy. The eye of the moral man is opened, and in the light of that new uncovering the knowledge of the glory of God is revealed in the face of Jesus.

(2). The other statement about saving faith must be as to its *object*. Dr. Hodge uncovers none. He tells us the object is Christ. He puts much of our believing in the form of trust. But Christ is not the object *ex differentia*, and Dr. Hodge smothers by Him the vital question. The object of the hypocrite is Christ. *Differentia* must be of such a form as to mark the Christ of the believer, and to mark the Christ of the impenitent man. *We* do that. We say, Faith is in a *holy* Saviour. And here it will be seen how the *nature* of faith and the *object* of faith can perfectly agree. Subjectively it is a moral act, and objectively it has a moral object. "We *know* what we worship" (Jo. iv. 22). But Dr. Hodge simply states a history,—'It has the testimony of the Spirit;' and when we hunger after that to find a *differentia* that shall have a distinguishable nature, he tells us that we know only the effects, and when we ask after the effects, we get only words that are entirely common as to the saint and as to the unbeliever.

We can only fasten on his negatives. Dr. Hodge certainly says (vol. i. 53), "Moral truth requires moral evidence, and 'the things of the Spirit' the demonstration of the Spirit." The apostle John seems to imply (Jo. iii. 6, 9; and v. 18) that we know the Spirit by His moral effects; but Dr. Hodge takes the negative. (Vol. iii. p. 88) "It is not correct to say that faith is assent founded on feeling." Vol. iii. p. 93, He denies "that men believe the truth because they love it." "Love [is not] the essence of saving faith" (vol. iii. 94). Sanctification is the effect of faith (vol. iii. 108) in such a sense that this latter precedes repentance (vol. iii. 41) as cause and effect. We are not justified by faith "as a pious disposition of the mind," but (now mark the utter scorn of any practicable *differentia*) by "faith of which Christ is the object" (vol. iii. 98). This boldly brings forth our difficulty, which is that Dr. Hodge really does not define faith; for specifically of what faith is not Christ the object? On the contrary, he seems to deny any subjective difference. "Everywhere in the Bible the fact that any one believes is referred not to his subjective state, but to the work of the Spirit on his heart" (vol. iii. 72). Faith does not include feeling (*ibid.* 49). Again, "If justification is sanctification, then it may be admitted that love has more to do with making men holy than faith considered as mere intellectual assent" (*ibid.* 95).

Now far be it from us to take the opposite of all these positions. It would not answer as language of our choice. But much farther be it, certainly, to affirm them as they stand. "Protestants," says Dr. Hodge, "are strenuous in denying that we are justified on account of love" (*ibid.* 95). We are not suited by saying that we are justified "*on account of*" any thing, unless it be the righteousness of Christ. The preposition would have to be unusually defined, and, moreover, is chosen polemically. Nor do we *care* to say that we are justified by love. It leaves out what we are to teach hereafter in respect to the pre-eminence of faith (p. 201). But these propositions serve one purpose: They distinctly give attitude to Dr. Hodge. When he says, "So far as the testimony is moral . . . the faith is temporary" (iii. p. 74); or when he says, "Faith is founded not on the moral evidence of the truth, but on the testimony of the Spirit," etc. (p. 68); or when he says, "The fact that any one believes is referred not to his subjective state, but to the work of the Spirit on his heart" (p. 72); we are not disposed to let the opposites of these be the forms in which to express our reasonings, but we do gather a great departure from the morality of the gospel.

When Dr. Hodge says, "If justification is sanctification then it may be admitted that love has more to do with making men holy than faith considered as mere intellectual exercise" (p. 95), I seize that sentence as in the grasp of a vice. Then "faith considered as mere intellectual exercise" is the faith we are grappling after. Dr. Hodge will hardly father it. Brahm will hardly say, This earth-pot is the great All-Eye. Brahm will run to his defence, and utter many things that I cannot think wrong. But then his drift remains. It may have been half blundered upon. Still it is here:—" MERE INTELLECTUAL EXERCISE." And if faith is assent ; if it is not of its essence holy; if its object is Christ ; if it is based upon authority ; if such a base and such an object belong also to the faith of the impenitent ; if, therefore, saving faith has the authority of the Spirit ; if, however, this is invisible, and can be known only by its effects ; and if, finally, it is still not moral, and these effects are not subjective, but the "work of the Spirit" (p. 72),—we are in Brahm's cycle of truth ; and we can at least fasten upon one result, —that whatever there is indistinct in the discrepancies of Dr. Hodge, there is at least one agreement,—that faith is none of your instances of love, and that saving confidence is not moral in its wakening vision.

And we learn more of this as we advance in his reasoning.

Suppose, for the sake of argument, that he is under prejudice. Suppose that the tempting which he shares with all the Church comes to him on the side of superstition. Suppose that he is a fetich-man, and that he offers Doctrinalism as twin of Ritualism;—Ritualism appealing to the very most genuine texts, would not Doctrinalism also to the tenderest revelations of the gospel?

Suppose the problem be to steal off with holiness. Would not Satan accomplish it in the most stealthy forms with a rear-guard of the very words of the Redeemer?

Accordingly, how is it?

Recollect, millions are at stake, and like Eve in Eden worlds hang upon a wrong interpretation,— what would be the style of argument by which men would inaugurate the error?

This writer expounds it by three indignant appeals :—

I. If faith is holy, then works save us, and the work of Christ is cast out of the account (iii. p. 170, *et passim*).

II. If faith is holy, we are justified by holiness; and how stands the doctrine that we are justified by faith? (iii. pp. 93, 172.)

III. Besides, if faith is holy, then what of its effects? one of its most signal sequences being that it has the effect of sanctifying the soul? (iii. p. 108.)

I. In respect to the first, though Christ be holy, why may not we be? Though holiness be a ground of pardon, why may it not be a condition? Because holiness in Christ is the ground of our redemption, why may He demand any condition at all, and not make one essence of that condition to consist in holiness?

Mark the extravagance of the system. I must be made holy. Without holiness no man shall see the Lord. Works must be holy: and without works the man is lost. But they must be produced by faith. Though all graces must be of their very essence holy, faith, the leader of the band, must be stripped of this vital beginning.

The thing is impossible.

And the argument is strained.

For if the sacrifice of Christ is not detracted from by the demand of a condition, and that condition imply trust or a dozen other qualities or parts, why may not one quality be holiness? Or, in other words, granting the premise that we have been obeyed for only by Christ, where is the sharp conclusion that He may not make *us* obey as the opening fruits, or the vital test, or, for all those who prefer the language, the instrumental means, of the soul's acceptance?

II. So of the second point. 'Faith saves us. And just as Christ's holiness covers all the ransom, so our faith covers all the ground of first condition. Where is the room for holiness?'

Now let us apply that to Transubstantiation. We are saved by eating. There can be no doubt about that, for we have the express Scripture (Jo. vi. 53). Where is the room for faith; for our Saviour, having done all the work, frames for us the fixed condition? Ay, but we are to eat in faith. No, says the Papist, you make void the grace of the sacrament.

Now will any one point out the difference between these two arguments? We say,—Faith, but a holy faith. Dr. Hodge would say,—Eating, but a believing eating. If I must eat believingly, and still there is no detriment to the Mass, why must I not believe lovingly, with yet no detriment to the gospel? The circumstances are most absolutely similar. Faith and the Mass fare each badly in the opposite hands. And what is the reason? Rome drives faith out of the Sacrament; and Dr. Hodge, holiness out of faith. Why? Because each claims his definition. Give Rome her definition of what regenerates a child, and we are vanquished if we introduce faith; and give Dr. Hodge his definition of saving faith, and we are vanquished if we introduce holiness.

III. And so of the third point.

The third point is, that faith produces holiness (iii. p. 41). And what is the conclusion from that? That faith producing holiness, the two things must be something different!—a clear *petitio* like the last. Joy produces holiness. Is, therefore, joy not holy? Again, joy produces joy. Is the first joy one thing, and the last joy different,—except that it may be different in degree? Let us go a little deeper. Holiness produces holiness. The first holiness, therefore, is unholy! Dr. Hodge must have his definition, and faith is not of its essence holy, or else neither of the three are points that he can lean to for his vindication.

On the contrary, though we be justified by faith, much Scripture would be a mystery if Dr. Hodge define the exercise. Let faith be an assent upon authority, so as to be sharp and narrow, and to exclude what is moral from the field, and how baffled would we be with dissentient testimonies (see Rom viii. 24; 2 Thess. ii. 10). But let faith be in its essence moral, and the kernel of it is in every grace. Divine reasons accentuate faith (see Heb. xi. 6), but never to the exclusion of holiness. We may be saved by anything, let it be only holy. Because,—did the Most High mock Cain when He said,—" If thou doest well shalt thou not be accepted?" The grace of holiness must enter any act to make it saving. If thou doest not well, apply at once to the Helper. "The Sin-Offering lieth at the door." That happens, therefore, which would seem most natural, that the Holy Ghost breaks away from harping only upon our confidence, and lifts up to the forefront some other exercise of holiness (Ex. xix. 5, and 1 Tim. iv. 16): the Holy Ghost this signifying, that faith is not faith to rescue a man till it is holy (Jas. ii. 20); and as holiness mounting into faith has mounted therefore into character, all graces, as they must equally possess it, are equal tests of the soul's salvation.

§ 2. *Dr. Hodge's Contradictions.*

Dr. Hodge, having marked with such very distinct utterances the exclusion of holiness from anything as of the essence of faith,* wakens the usual surprise by sentences entirely opposite. He attacks Bellarmine, and says,—he "makes love the essence of faith" (vol. iii. p. 94). "In other words," he says, "love with them (i. e., the Romanists) is the form of faith: it is that which gives it being or character as a Christian virtue or grace" (*ibid*). He says, "all this is intelligible and reasonable provided we admit subjective justification and the merit of good works" (*ibid.*). And he joins with Luther in his violent language,—"This pestilent and satanic gloss I am not able in any temperate way to detest" (*ibid.*). And yet he has already said, far back on the fiftieth page,—"The perception of beauty is of necessity connected with the feeling of delight. Assent to moral truth involves the feeling of moral approbation. IN LIKE MANNER SPIRITUAL DISCERNMENT (FAITH WHEN THE FRUIT OF THE SPIRIT)

---

* It is astonishing how this pervades our literature :—"The doctrine of justification by faith alone is so simple and self-evident to all Protestants, it has been so inculcated from the cradle, that it is difficult for us to realize how it struck the pious monks in the days of the Reformation. For it is a doctrine which the Church of Rome has ever stamped as heresy. Rome teaches that men are saved, not by simple trust in the free mercy of God through Jesus Christ our Saviour, but by love and good works. We are pardoned according to Roman Catholic theology only when we become good. According to Protestant theology we become good, only after we are pardoned." *Christian Weekly, March* 14, 1874, p. 125.

INCLUDES DELIGHT IN THE THINGS OF THE SPIRIT, NOT ONLY AS TRUE, BUT AS BEAUTIFUL AND GOOD. This is the difference between a living and dead faith. This is the portion of truth involved in the Romish doctrine of a formed and unformed faith" (iii. p. 50).

Is not this startling?

The sentence we have picked out in capitals takes up all our doctrine in the case to the uttermost crumb. Dr. Hodge joins us at a leap. Look all through the sentence. "Delight in the things of the Spirit as beautiful and good:" what is that? Delight in Christ, for example, as good? If words can say anything that is more thoroughly an account of holiness, we are unable to select them. And yet mark! he does not speak now of a consequence of faith, but of what it "includes." And then the very word in its English dress,—"*the difference.*" The *differentia* "between a living and dead faith" is that it "includes delight in the things of the Spirit, not only as true but as beautiful and good."

But let us go on to other cases.

On the eighty-ninth page Dr. Hodge says,— "His faith is founded on the inward testimony of the Spirit by which the glory of God in the face of Jesus Christ is revealed to him." And on the ninety-first page,—"This INCLUDES the apprehension and conviction of His divine glory, and the adoring reverence, love, confidence and submission, which are due to God alone."

We had a hand raised at the opening of our treatise to erase a sentence in which we spoke of contradiction as though it were a mannerism of Dr. Hodge; but really was it unjust? "A delight in things as good" is not a bad definition of holiness; undoubtedly in the broad philosophies nothing can be more discriminatedly holy. Dr. Hodge declares that it is the *differentia* of what saves. And yet it is after this that his expulsion of it is so abrupt; and as late as the ninety-fifth page, that he sheers away everything but "mere intellectual assent," and says,—" If justification is sanctification; " that is, if by making faith moral all things else must be subverted; " then it may be admitted that love has more to do with making men holy than faith considered as mere intellectual assent." Let us finish the passage. "And if it be conceded that we are accepted by God on the ground of our own virtue, then it may be granted that love is more valuable than any mere exercise of the intellect."

§ 3. *Argument from Scripture.*

The texts that Dr. Hodge quotes are like the scooped rinds on All-Hallows Eve, that get their features from the fire that we put within them. Give Dr. Hodge his definition of faith, and, of course, holiness is not essential to it; but then the text simply serves as a frame, and the whole logical result is arrived at in the preliminary definition.

So then let us dispose of the texts in blocks, like the great bankers on the Stock Exchange.

First, all those texts that speak of faith as the whole of that by which we are justified. Of course holiness is an intruder upon the work, unless it be of the very essence of faith. But Dr. Hodge never gives us the benefit of our own frankly stated supposition. We believe that color is in its very essence light. If, therefore, we admit that all things are made visible by color, he is enforcing upon us his own different definition who says that they are, therefore, not on this very account also made visible by light.

So also in all the other texts. There is a *petitio*. We grant the premise, but we are always put to the wall by a forced conclusion.

For example, another block of texts. "By the deeds of the law shall no flesh be justified" (Rom. iii. 20). A man is a Deist who does not take all those asseverations. But what do they distinctly mean? They distinctly mean that Christ obeyed the law, and wrought for us a complete and all comprehending atonement. But does that mean that He believed for us? Did He work out in His heart, or did He work out in ours, the all-needed condition of a subjective confidence? If He wrought it out in ours, then you are putting your definition into faith, if you say it is not a holy faith. We must have a holy hope, and a holy love, and a holy joy, and without every one of these things we cannot see the Kingdom of Heaven. We must have a holy holiness, and confessedly without holiness no man can see the Lord. And if things so express as this do not damage the obedience of Christ, why a holy faith?

Then another block of texts. Faith specially interferes because it is the great entrance into grace. It cannot be moral because it wins morality.* It cannot be of its essence holy because it is the path to holiness (Acts xv. 9). It is the hand reached out for our redemption (Col. ii. 6). But now, rigidly, we demand a choice. Is it or is it not un-moral because it interferes with Christ? Is it or is it not un-moral because it makes men moral? Let us bind the argument to the intended proof. It is not un-moral because of the obedience of Christ, for His perfect obedience does not forbid the grace of obedience as the condition of acceptance. Nor is it un-moral because it makes men moral (Acts xv. 9). To show that faith and holiness must be apart, we must show in this class of texts that holiness itself does not produce holiness among men (see above, p. 187).

But now, turning the tables, what is meant by our being "justified by works?" (Jas. ii. 24.) Isaiah tells us to "wash us, make us clean," and then immediately adds, "Though your sins be as scarlet they shall be as white as snow." Ezekiel says, "Make you a new heart" (Ezek. xviii. 31). The Saviour preaches, "Repentance and remission of sins" (Luke xxiv. 47), and Peter repeats,—"Repent and be converted, that your sins may be forgiven you" (Acts iii. 19). Matthew never speaks of faith until his

* We use morality in the philosophic sense.

sixth chapter. John thunders out, "Repent," and the axe is laid for them that bring not forth good fruit: and Jesus, in His only extended speech, fills out all its paragraphs to the very end about a righteousness which is not His righteousness at all, but a righteousness that must be greater than of the Scribes and Pharisees, and must be a righteousness such that if a man hears about it and *does it not* he builds his house upon the sand (Matt. vii. 26).

Now, theologians mock the Scriptures when they say that James means not that a man is justified but that his faith is. Burkitt and Henry both use this wanton arguing. Because, where may it stop? James deliberately approaches an abuse in dogmatic theology. Paul had been preaching the doctrine of justification by faith. Holiness had been swept out of it, just as Dr. Hodge has taught it. James takes fire at the mistake. Undoubtedly we are justified by faith, and Paul beautifully taught all that transcendent gospel. Undoubtedly we are justified by grace. But now, only supply a faith that is of its essence moral, and all beautifully falls into its accepted place. James does not say our faith is justified by works, but "a man is." "Ye see then how that by works a man is justified and not by faith only" (Jas. ii. 22), and in James all binds itself beautifully together. Repent and be saved (Acts iii. 19); why, if we are justified by faith? Because we will not believe unless we repent. Why then not say,—Repent, and no longer say, Believe?

Because we will not repent unless we believe. Each is wrapped up, and each inside the one of the other. Hence we are justified by works, because holiness is of the very essence of faith, and faith never justifies without this sole *differentia*, the very highest in its nature. Repent and be saved. Why? Because the moment a man is holy he is saved. Believe and be saved, because a man is never holy till he believes; and it has pleased God, as we shall afterwards show, that for the honor of Christ, and as His intended scheme, it shall be a vision of His excellence that shall be the first holy act that shall unite us to the Redeemer.

Would it be safe to give up faith and try to be holy to get into the kingdom? I say, No. Not that a holy act would not be entrance there; not that a holy act would not prove that we had been admitted; but that a holy act will not be given in the neglect of Christ; it has pleased the Father that Christ shall give it; it has become the plan that we shall seek holiness at the hand of Christ; and hence we dawn into being holy in an open eye upon the excellence of the Redeemer.

§ 4. *Dr. Hodge's Argument from Reason.*

If our *differentia* were granted, namely holiness, it would leave a wholesome liberty for all the other characters of faith. Philologically considered there is no end to its variety. I believe anything,—sure or radically doubtful,—from a ghost story that the most credulous might reject, up to self-evident truth like the testimony of sense. Psychologically we are just as wide. Faith really is no special psychologic term. Nor in religion has it, outside of grace, any psychologic fixity.

Dr. Hodge, however, had a certain head to bore to. Like Shanly in the Hoosac mine his headings must come out together. He begins, therefore, back among the hills; and as faith that saves is to be belief on the testimony of the Spirit (iii. p. 68), he gives that heading to the work away back in the region of psychology. He defines about thus:— Psychologic faith, i. e., belief upon testimony; religious faith, i. e., belief upon Authority; saving faith, i. e., belief in Christ on the testimony of the Spirit. When the blast is heard, day-light passes all through the bore from one end to the other.

But really, inspect this work. Is faith a term of psychology? Or if it is, is it not a random term for any form of belief or conviction? Why does Dr. Hodge say it is not a belief in this thing or that thing?—in the unseen (iii. p. 53), as Lombard expresses it? in the un-positive or what is only probable? (p. 46.) Where does he get any restriction for the use of faith? Or when we arrive into religion, why may not things win my faith when they suit my conscience, or when they tally with my reason? Why must I always believe upon authority? particularly when the Bible tells me that "the invisible things of Him from the creation of the world . . . . are understood by the things that are made?" (Rom. i. 20). Where, logically and by laws that would be exacted in science, does Dr. Hodge get the right to chain a word down to unity when it is so common as faith, and thus have it all ready to shed its probabilities upon that faith that he describes as saving?

And even afterwards, when the headings meet, what use can he make of his tunnel after all? What knowledge does it bring to us of the actual *differentia* of faith? He says, it "is founded on the Spirit's testimony" (iii. p. 68). Suppose we grant it. Suppose we agree with Dr. Hodge (and nothing can be easier historically) that it is the Spirit that brings the faith, and that we never could enjoy it without the Spirit's testimony. That is all so. But what is all that but an invisible transaction? Do we taste it? Do we see it? We taste the fruits of it. And there is the very question. What are the fruits of the Spirit's testimony? Dr. Hodge distinctly buries this. What is the *differentia* when the Spirit works? We say, holiness. Dr. Hodge admits, "When a man . . . believes he is conscious only of his own exercises" (p. 88). He avows, he "is not [conscious] of the supernatural influences of the Spirit" (p. 89). To these "those exercises [only] owe their origin and nature" (p. 89). Dr. Hodge, therefore, has not given a conscious *differentia*. And when we attempt one and say, The *differentia* of faith is a sight of excellence granted by the Spirit, he turns his back upon it. As though I were to say,—The *differentia* of art is beauty. Dr. Hodge might retort, No; the *differentia* of art is genius. I might respond, True enough:

that is the historic account of its creation: but the *differentia* of art, as an apprehended verity, is the beauty of the work, and genius is the unseen skill that attests itself and exerts itself in its execution.

We will not decide for Dr. Hodge, but sometimes he seems to doubt a *differentia*. "Every where in the Bible," such is his distinct avowal, "the fact that any one believes is referred not to his subjective state, but to the work of the Spirit on his heart" (p. 72). Then there is no subjective difference. Then the psychologic faith was the saving faith, with this only interposed, that this last is the work of the Spirit. Then subjectively, and so consciously, there is no apprehended difference. Then psychologically defined faith is hermeneutically defined and dogmatically defined and soteriologically defined sufficiently and all in one. Then faith very rightfully is "intellectual assent" (p. 95); and Dr. Hodge so glorifies trust, that, grant the object is Christ, and grant the subject is the lost, and grant the giver is the Spirit, then faith has no conscious mark at all. It is only intelligent assent to the things of the gospel.

On no other foot can we understand certain very grave asseverances. "Faith is [not] founded on feeling, because it is only of certain forms or exercises of faith that this can even plausibly be said" (p. 89; see also pp. 52, 60). Again. "There are many exercises of even saving faith .... which are not attended by feeling" (p. 89). And again,—

"This is the case when the object of faith is some historic fact" (*ibid.*). Let us lay stress on this.

Faith is either a habit or an act. In other words the expression "*faith*" means either the grace of believing, or the saving act by which a soul is reconciled to God. In our theory the discrepancy is a matter of indifference. In either meaning holiness is the *differentia* of the faith. But with Dr. Hodge what are we to understand? If faith be a Christian grace, then are there some forms of Christian grace that do not include feeling? Or if faith be a saving act, then are there some saving acts that include it,—or not, as the case may be?

The *differentia*, then, must be in the intellect. The psychological beginning-back was not, therefore, so far out of the way as we might imagine. Faith is a belief upon testimony (iii. p. 60). Sometimes it is moral. Sometimes it is not (i. p. 53). Its being moral is not therefore the *differentia*. The *differentia* is the testimony of the Spirit (iii. p. 68). And as the Spirit is invisible, and we are conscious only of our own exercises at the time (iii. p. 88), there is nothing subjective in the case (p. 72), unless it be an intellectual belief (p. 95), and those after gifts which are the reward of our believing.

§ 5. *Our Own Doctrine.*

Holiness enters in two ways into the work of our salvation. Perfectly, it became the ground, germinatingly, it became the test, of the sinner's acceptance with the Father.

A priori there is no clashing, that Dr. Hodge ever points out, of one of these holinesses with the other.

Perfectly, holiness in Christ obeyed the law, and wrought out a perfect ransom for the relief of His people.

This relief might have been immediate.

It has pleased the Father that it should not be, in three important particulars.

To wake in heaven after having slept under the frown of the Master would not be beyond His grace, but is aside from His ordination.

I. It has pleased Him, first, that we shall become holy in this world. Hence, before we go any further we say,—If any man shall become holy he shall be saved, and that as an iron promise all through this blessed revelation (Is. lv. 7; Mic. vi. 8).

II. But it has pleased Him, second, that we shall hear the gospel. Not only are we lost and dead, and germinating holiness must be imperatively His gift, but He chooses to give it exclusively by the word.

See now how far we have got. He chooses to turn us by a struggle in ourselves. He chooses to germinate before He renders complete. He chooses to do it formally in every instance of adult life, and to do it through a struggle in ourselves, and to do it till we become holy, and to make this by statute ordinance the test; and therefore the promises that he that will cease to do evil shall have his sins white as snow (Is. i. 18).

III. But then, thirdly; not only does He give this holiness by the word ; that is, not only does He print upon our hearts the law of holiness and use as His instrument the letter of an understood redemption,—but, as an equally inexorable fact, He demands of us an acknowledgment of the Redeemer.

Let us be perfectly understood now. We do not recall the other. We do not recant that whoever repents shall certainly be forgiven (Luke xxiv. 47). But we state two modifying facts, first, that we will not repent without the word, and secondly, that we will not be helped to do it except by calling upon the blessed Redeemer.

Ho, then, we are at the side of Dr. Hodge !

No ; for there is the first demand. We are to become holy. " Without holiness no man shall see the Lord " (Heb. xii. 14).

We might have got it in our sleep. We might have learned it in the ten commandments. We might have caught it in our drink. There is no limit to God's power. But it has pleased Him "by the foolishness of preaching" (1 Cor. i. 21), and then also by our seeking to Christ, to give the power that lifts us to a beginning of salvation.

What then is the history? The sinner perishes. What must he do? He must inexorably repent? How can he repent? Only by God's help. How is that help given? Through the word. Does the word always convert? No. What is the difficulty? He must cry out to Christ. This now is the circle of the history. I will never be saved unless I repent; and I will never repent unless I seek it of Christ; and faith is the pregnant word which decribes the state in which I find my repentance.

A man impenitent is ever growing worse. A man converted is ever growing better. To keep on growing better I must begin to do so. To begin to grow better I must have the help of Christ. To have the help of Christ I must ask it. And to ask it is the beginning of that common faith that ends at last in the faith that is holy.

Here are the "Jersey" and the "Central;"— here are faith and repentance. I tell the man to repent, and he says he cannot do it. I tell the man to believe, and he cannot do it. I tell him to repent, but tell him he never will repent unless he believes. I tell him to believe, but that he never will believe unless he repents. I tell him that either, and therefore, both, are necessary to his forgiveness. If he falls back upon me in his despair, I tell him more plainly still that it has pleased God that He must repent, but that he cannot repent without getting it from Christ, and there instantly, is our need to ask Him; that He moves at once at our entreaty, that He quickens even our common faith, that He kindles our zeal to ask and moves us nearer. And if you say, Why do you tell me to repent? I answer, Because you may be trying that too. It will all bring you nearer. "The law was our school-master" (Gal. iii. 24). Repentance means *thinking-after* (*nachdenken*), i. e., remembering our sins. That will all help. Meanwhile push your plea with Christ. I say, it is perfectly simple that the two rail-tracks become finally one. They come together when you reach the platform. Look under the train. The "Jersey" and "the Central" have become a single rail. You pushed repentance. That helped. You cried out believingly. God helped your unbelief. There came at last the moral vision. The two graces married into one. Faith bloomed at length into what was holy : and only then it prehended Christ, and *ipso actu* became repentance.

In the Repertory of 1842, Dr. Hodge writes as follows :—" Truth and holiness, error and unholiness are so inseparable as to be rather identical than united. They are different phases of the same thing. What is light to the eye is warmth to the hand. What the mind apprehends to be true the heart feels to be good. Hence in Scripture the word truth often stands for moral excellence, and the belief of the truth includes the love of holiness" (p. 143, Rep. 1842.)

On the contrary, reviewing Mr. Barnes, he rejects this whole moral view as not at all the critical one :—" Mr. Barnes in his subsequent remarks says expressly, ' Faith is not the meritorious ground of acceptance, for then it would have been a work. Faith was as much his own act, as any act of obedience to the law.' And again, ' Faith is a mere instrument, a *sine qua non*, that which God has been pleased to appoint as a condition on which men may be treated as righteous.' This is all very good, but he immediately turns the whole matter round when he proceeds. 'It expresses a state of mind which is demonstrative of love to God; of affection for His cause and character; of reconciliation and friendship; and is THEREFORE that state which He has been graciously pleased to promise pardon and acceptance.' This gives a sadly erroneous view of the relation of faith to justification. Faith is the instrumental cause of justification because it is the means of our becoming interested in the merit of Christ, and not because it is indicative of love to Christ, or of reconciliation or friendship."

"Faith is no more demonstrative of love to God than repentance, gratitude, self-denial or any other holy exercise" (Princeton Review, 1835, pp. 297, 298).

This last nobody denies, but it is really the key to the orthodoxy of Mr. Barnes (I mean in this one case), and to the heresy of his great reviewer. These all do possess a common essence, and therefore it is that they are all saving. A man is justified at once if he possesses one of them; nay, never justified till he possesses all of them. And in very truth he began with each of them, for he possessed them all when he put his hand upon his blessed Redeemer.

§ 6. *Anything else Fetich.*

Fetichism, when we come to fathom it, is not only a thing which has long ago dispensed with holiness, but which, as a judicial consequence, has dispensed with reason, and as an eminent feature in this last, has dispensed with reason in a bare and baseless trusting to authority.

I. Dr. Hodge sells out to Science in his definition of *common faith*.

I know of nothing more forlorn in sacred literature.

The Roman Catholic does not leave himself without a bottom. He tells us of the authority of the Church; but foreseeing the difficult retort that authority without base is fetich, he carefully furnishes one. He tells us that we are to judge beforehand *why* we are to obey the Church (Co. of Trent); and that we are to judge, besides, *what* we are to obey, viz., to inspect the Church, and only then to give it our eternal confidence. The Roman Catholic system, therefore, is consistent with itself. But Dr. Hodge talks like Boodh or Vishnoo. We are to believe solely on authority (iii. p. 67); and though he falls upon this thought of *judging* what is true (i. p. 51), and goes so far as to give us a *judicium contradictionis* (i. p. 52); yet this is one of his recessions. He soon goes on to the extreme, that we are to believe, "not on rational or philosophic grounds, but upon the authority of God" (iii. p. 65). Hence we are to have a faith "not founded on sense, reason or feeling, but on the authority of Him by whom it is authenticated" (p. 63). And then, cutting all away, we are hung like Mohammed's coffin :—" Even our assurance of the veracity of consciousness is of the nature of faith" (*ibid.* p. 48); and, at last, putting reason quite out of the field, and making a burlesque of what was admitted of reason as a *judicium contradictionis*, we read this:—" The original data of reason do not rest on reason, but are necessarily accepted by reason on the authority of what is beyond itself" (*ibid.*). We shall return to this (p. 217).

But now, what chance have we even with this view of *common* faith with such men as Huxley? They have the victory in the outset. Snuffing out all light of reason, they can say what they please. Hence the talk that has become so common, that religion is in one sphere and science in another. Hence the apology that is so shamefully pushed, that religion was never meant to teach science. Hence the inroad, like that upon the Roman Republic, by which the fasces of the tyrant are brought into the very Senate Chamber of the people, before they know that their liberties are thoroughly overthrown.

Give the Huxleyans all the empire of reason, and give theology nothing but that word authority, and —still our Grand Head will secure to us the victory; —yet it will only be by judgments upon the Church; by crushing up this potsherd of implicit belief; by scourging the followers of this false faith; by rehabilitating the stewardship of reason; and by suffering years to pass with the strong intellects of the world debauched, and our abandoned gifts employed in crushing our Redeemer.

For now think of an idolater. Here is a gree-gree. Why do I wear it? Because I am not holy.

Give us a more proximate account. Because I am not reasonable. Give us one nearer yet. I wear it on authority. I wear it more definitely than that on the authority of my Deity. But who is this Deity? Ah, there you approach my faith. I say, the man who believes in the authority of God, and cuts loose from faith in an entire supremacy of reason, is believing on no authority whatever. He is not honoring God *in esse.* He is believing in no God. The moment we cease to "know what we worship" (and that knowledge must be a thing of mind, moving as in other things), we cease to have anything to worship, and we have no right then to undertake the defence of Christ,—or anything.

For now more definitely yet : why do I believe? I believe on authority. Why do I believe on authority? Simply on authority. Why, that? Still on authority. The very belief in consciousness stands at last upon authority. And Jones from behind the plow can come in, and have all that explained, and tell you in an instant,—My dear sir, you have no authority at all. You have rolled up everything into a ball; and unless you can snatch back reason from the scientists, you have sold the truth to its most hungry and persevering persecutors.

II. But if this is the attitude of Dr. Hodge's *common* faith, how must it be with his faith as *saving?*

I. In the first place, it *bewilders the impenitent.*

His *common faith* is a belief on authority. His *saving faith* is a belief on the testimony of the Spirit. In either case the testimony is out of sight, and in neither case will he suffer us to bring in sight the scheme of a moral essence.

I, following the sinner in the very foot-prints of his sin, tell him he must repent. I never leave him for a moment off the scent of his iniquity. I tell him he is a lost wilful sinner, and that his helplessness consists in his disinclination. I tell him that Christ has died and that all righteousness and all power to change his nature has been purchased for him by the cross of the Redeemer. I tell him, therefore, that his nature must be changed, and that that change in this world becomes the signal and the test of his justification with the Father. I tell him that it has pleased God not only to change him in this world but to do it with his own efforts,—to do it with the teaching of the word, and to do it, thirdly, in an acknowledgment of Christ, and that faith, therefore, which is a word that wraps all that up, is the necessary act for the soul's acceptance.

I tell him, therefore, to speak to Christ; and, as the essence of what I define as faith is simply moral, I am able to keep his mind clear by a distinct *differentia*. I tell him that what he wants is holiness. I tell him, it makes no difference how that holiness appears, it is the test of his salvation. It may be hope (Rom. viii. 24); it may be love (Jo. xiv. 23); it may be wisdom as the old Faith called it (Prov. iii. 13); it may be zeal (Num. xxv. 11–13);

it may be good works (Jas. ii. 21); it may be penitence (2 Sam. xii. 13); it matters never a word. It may be anything that will bear the moral test (Ec. xii. 13). Only, God is resolute that it must dawn forth when the creature is seeking to Christ; and, therefore, it must be faith what time it gets this holy light, and what time it has born into itself all these moral graces.

This is in every part simple to the sinner. And I never attempt pressing him by a single thought that does not keep forcing upon him the need of his repentance.

In fact, I give him no discouragement. I do not tell him he must not work. I do not say as a preacher did the other day, that the cry, "What must I do?" was replied to categorically by Paul that virtually he must do nothing. I call upon him by all that is sacred to do all of ten thousand things that will help in the least toward a change of heart. But then I tell him, You never will be moved to renounce your wickedness except at the call of Christ; and, therefore, keep calling to Him. Use all these means in earnest appeal to the interposition of the Sanctified One.

What do I do, therefore?

I allow the sinner to be saved by anything that is holy. I show him that that is the way the Bible speaks of all holy attitudes of mind and virtues of the heart. I lead him to feel that holiness is a unitary state; only that it is gotten first by seeking; that it is peremptorily taught that it must be sought from Christ; and, therefore, that this duty, which must be holy like any grace, is naturally that duty of the lost which first has birth in this moral excellency.

Now here is no wavering.

But look at the other view.

1. I teach a *sinner* with it.

He asks me what it is to believe. I tell him it is to trust. He asks me what he is to trust. I tell him, the gospel. He replies, he does trust the gospel: he has been taught all it utters from his mother's knee. I come to a point where he obliges me to make some distinction. I tell him it is on the testimony of the Spirit; but he soon burrows into that, and obliges me to say that all this supernatural history of the change is invisible, and can be known only by its effects. I am confronted, therefore, by a need of a test.

I tell him it must be a living faith, and that it must be FOLLOWED by holiness. He begs,—Just one thing at a time! He is concerned *a parte ante*, and begs some clear *differentia* of the *act*, inasmuch as what it must be *followed* by is a thing that concerns him after it has been achieved.

Of course it bewilders him.

To do nothing, to feel nothing, to be very careful not to work, to be very certain only to believe, and to have faith intruded in a form in which it must now appear that theologians cannot discriminate themselves, is terrible either for the unfeeling or the convicted.

The unfeeling will put it on the shelf. In the first place they are commanded not to work, and in the second they are instructed in a superficial remedy. The convicted, puzzled where they sought the light, and giddy by an indistinguishable difference, are tided over the time of hope, and made hypocrites of by a fetich-believing.

2. For, secondly, this is another charge. It *makes hypocrites.*

Dr. Hodge has no chapter in all his work on the subject of Repentance. This is a startling fact.

If men are distinctly taught that repentance *follows* that act by which alone they can get an interest in Christ, repentance will be kept out of sight, and they will school their views upon a belief that is not moral.

3. Hence, thirdly, less thought about the duty of *moral training.* Men will prefer to "love much" (Luke vii. 47). Men will talk about the danger of the moralist. Repentance, not being in the very eye of faith, need less to be prepared for by years of tutelage. And hence that modern church,—all feverish for faith,—all clamorous for immediate results,—credulous on the gibbet and in the jail,—but not breathing courage on the mother in her patience and moral care ; so that the revival of the church, and the outpouring of the Spirit in the sanctuary, become suspected of men as separated from a training for redemption.

4. Lastly, it lets down the whole tone of exemplary profession. Beginning unmorally, men end

that way. Tell a man that he must be born again, and that the faith that saves him must go down to the bottom of that work, and he will not want to profess it till it has something real.

But tell him he must wait for all this,—that he must believe, and then look for repentance,—and he will maunder at first as having no very distinct idea of such a believing, but, waking up as we sometimes hear it expressed to the simplicity of this act of trust, he may do nothing more than he has always done since he was first taught about Christ, and never trust with any possible repentance.

We are never pardoned except we are born again. We are never born again except in holiness. This is the very essence of the work. We are never visited by this change unless as we are moved to seek for it from Christ. This seeking never mounts up to a justifying faith till the regenerating act, and till faith like birth is a sight of holiness.

CHAPTER X.

RATIONALISM AN OVER-USE OF REASON.

§ 1. *Dr. Hodge's Statement of his own Doctrine.*

"By Rationalism is meant the system or theory which assigns undue authority to reason in matters of religion" (vol. i. p. 34).

§ 2. *Dr. Hodge's Exposition of his own Statement.*

"By reason is not to be understood the Logos

as revealed in man, as held by some of the Fathers, and by Cousin and other modern philosophers ; nor the intuitional faculty as distinguished from the understanding or the discursive faculty. The word is taken in its ordinary sense for the cognitive faculty, that which perceives, compares, judges, and infers " (vol. i. p. 34).

It is an advantage to start with this degree of precision, for there is some ambiguity in the word *reason*, and to be stopped in the midst of a debate by the cavil that *reason* means *reasoning*, would not indeed affect the argument, for *reasoning* is just as wide as *reason*, and "an undue authority" of one would exactly bound an undue authority of the other; but it would take time to show that. Dr. Hodge saves us from all such task, and tells us that the word is " in its ordinary sense." The doctrine therefore that Rationalism is an undue authority of reason, means an undue authority of " the cognitive faculty," viz., " that which perceives and judges."

And how it is an "undue authority" Dr. Hodge goes on to explain. " In making the reception of Christian doctrine to rest on reason, and not on authority, the Dogmatist and the Rationalist are on common ground" (vol. i. p. 46). There are no parts of his book in which the shell-like or myth-like doctrinalism of Dr. Hodge more distinctly appears. Then there are truths that are founded exclusively upon authority, and are not amenable at the bar of reason ? So a Hottentot might defend his charms ; and so the Roman Catholic Church, with the same baseless self-forgetfulness, cuts off much private judgment from the people. In fact Dr. Hodge uses the same argument,—"The poor cannot be Rationalists" (p. 41). That is to say, unless faith is founded on testimony and not on reason, only philosophers could be truly religious. We could almost fancy that it is Bellarmine defending the testimony of the Church. Again, "Rationalism assumes that human intelligence is the measure of all truth" (vol. i. p. 41). And as Dr. Hodge discards this, he tells us what he puts in its place. "Faith is assent to truth founded on testimony; '*credo quod non video.*' Knowledge is assent founded on the direct or indirect, the intuitive or discursive, apprehension of its object" (p. 41). And then he tells us what the great question is;—"The great question has ever been, whether we are to receive truth on authority, or only upon rational evidence" (iii. p. 62). Lest any one should say,—Dr. Hodge is only awkward in delivering his meaning; he means that we are to believe as we believe in gravitation, unable to comprehend any of the facts, but, on the basis of general evidence: Dr. Hodge puts it out of our power to administer any such palliation. He does indeed contradict himself (see p. 41) in this immediate place. He tells us, "Faith implies knowledge;" though on the very same page he has set the two as opposites, building "faith" "on testimony," and building "knowledge" on the "apprehension of its object." Or take another instance. "Faith implies knowledge;" but "our duty, privilege and security are in believing, not in knowing" (p. 48). He tells us, "rationalism assumes that human intelligence is the measure of all truth" (p. 41); and then, after naked sentences which declare that testimony is all this, he rolls back, without the least apparent sense of inconsistency, to sentences like these;— " The indispensable office of reason in matters of faith is the cognition or intelligent apprehension of the truths proposed for our reception" (47); and again, —" Knowledge is essential to faith." This is the wabble of the machine-pivots, which, when they play to one side, play back, *ex necessitate*, too much to the other. Error cannot be steady. Dr. Hodge means to stand by testimony, and, to show that he does, he rids us of all doubt by saying, that testimony is the ultimate support even in reason (iii. p. 48). How gross an affirmation! (see p. 217.) Nay, Testimony is the ultimate support even in consciousness!! (*ibid.*) What an idol-God is such a Jehovah!

And then the likeness to the arrogations of the Papacy! The faithful must believe on testimony! No matter, now, why we are endued with reason. No matter how it is the voice of the Almighty. No matter where we are to be held responsible for its workings. Our very consciousness, which Dr. Hodge admits must be the last appeal, is so not the last appeal that it is founded itself on the testimony of God. What must the Huxleyans think if they dissect with care all this unthrifty reasoning?

§ 3. *The Doctrine False.*

Reason, as Dr. Hodge admits, is "the cognitive faculty" (i. p. 34). As we have just seen, he makes it as wide as consciousness (p. 49). This is a glorious sense: and beyond a cavil it is the only true one. Reason is the whole mind in its perceptive or intelligizing aspect. Now what exercise of consciousness does not come within the periphery of such an attribute? I see a picture. All its harder traits are under the eye of reason. I feel a picture. That is, I see its exquisite charm. Is not that reason? Why, I can turn and take it up in the forceps of my thought, and reason upon it like any other thing. Who tells me that cold and heat may be subjects for my *nous*, and that beauty and gentle tastes can be nothing of the kind? So of holiness. There is nothing apprehensible by the mind, or to be introspected by it, that is not rational. All things, therefore, appeal to reason, and it is the final arbiter. It is that which we shall have to give account for in the day of judgment.

And there are three stages of the error: first, that which supposes that there are some things that belong to faith, and some things that are within the province of reason. All things are within the province of reason. Faith itself is an act of reason. Reason is the mind in one of its aspects. And there is no healthier sort than that form of intellection that counts all worship as an act of reason, and this as the broader term that includes all the acts for which we are responsible.

If any one asks, Is this the old way of speaking? Possibly not; but this only shows how the fetich-unrationalness has grown upon the Church.

But if any one asks, May it not be a logomachy, therefore? may it not be just a question of width in the term of reason? We say, No, most signally. For what has Dr. Hodge declared? That reason is based upon authority.

Let us recur to his very words. "The ultimate ground of faith and knowledge is confidence in God" (vol. i. p. 52). Here is the same denial of truth *in se*, and the same appeal to a mere Supremacy (p. 406). Again, "Even our assurance of the veracity of consciousness is of the nature of faith" (vol. iii. p. 48). And again, "Reason itself must at last rest upon authority" (*ibid.*). Then what a foot-ball it is. We confidently aver that this bandying of reason is a sign of the region of idolatry. "Because they did not like, etc., God gave them up," etc. For observe, what is Dr. Hodge's very last appeal? Incontestibly, reason. He is forced to call in that stalwart power at the last extreme of doctrinal rejection. For listen, "The Scriptures never demand faith except on the ground of adequate evidence" (vol. i. p. 53). "God requires nothing irrational of His rational creatures. He does not require faith without knowledge, nor faith in the impossible, nor faith without evidence" (p. 55). And again,—" That reason has the prerogative of the *judicium contradictionis* is plain from the very nature of the case. It is a contradiction to say that the mind can affirm that to be true which it sees cannot by possibility be true. From the very constitution of our nature we are forbidden to believe the impossible" (p. 52). How unhappy to represent the Church in the most important recent theologizing by saying immediately after this, " The ultimate ground of faith and knowledge is confidence in God."

But more unhappy, perhaps, the intrinsic nature of the doctrine. Here will be our second stage.

What is authority?

Dr. Hodge will say, The authority of God.

But look ye here,—that *in principio* is ink and calf-skin.

What makes it a divine authority?

In the first place, Who is God? In the second place, What cares He for this Book? Dr. Hodge utterly forgets the decisiveness of reason in picking out all the canon.

Then, as a third stage,—How do I know the obligation of authority?

It is easy for Dr. Hodge to say, Not reason, but authority. But what is authority? The pilgrim casts himself under the car of Juggernaut in a trust to authority. Is that the sort of authority that Dr. Hodge would plead? The Papist eats the flesh of Christ on the faith of authority. Is that sufficient? Does not authority itself stand in judgment? And are we not to settle its claims? And is it not a sin in all these heathen tribes to do the very thing that Dr. Hodge appoints, viz., to build our faith, not upon reason, but upon the dictates of a mere authority?

To say that the sense is that we must believe doctrines without understanding them, is only to stuff rubbish in our works, for what does any angel believe except what he does not understand?

Dr. Hodge ought not to waver in that way.

Is or is not the gospel rational? If not, then we ought to give it up, for Dr. Hodge himself says we are not bound to believe the irrational (vol. i. 55). If it is, we ought to run up our flag to the peak. Religion can hold her own before the world. We ought to crush the scientists by answering them. Not always in each particular case (for we do not hold with Dr. Hodge that we are at liberty to disbelieve what contradicts reason) but by the force of a general aggregate. If ninety out of a hundred points are in favor of religion, I wait for the ten. If the vast volume of the proof is rational and on the side of Christ, I bear with a trifle even if it contradict my senses.

§ 4. *The Mischiefs of Believing Otherwise.*

We live in an age when infidelity is firing all her volleys against the truth. It is not Science: it is Satan. We agree with Paul, that men are "contentious, and do not obey the truth," but with the baldest prejudice actually prefer to "obey unrighteousness;" and, therefore, we observe them with the greatest eagerness obeying the theory of to-day, in preference to the splendid system of an established gospel. What does the Church do? Why, many times, as in the instance of Galileo, nourish infidels by a contempt of reason; many times, as in the instance of Voltaire, nourish infidels by an absurd authority.

The same hand-writing appears in our day.

We utterly protest against any syllable in Dogmatic Theology that sets up authority bare of reason. It is fetich. It is God, no God. It is authority with no shadow of a foot. It is a barring out, where there should have been the very finest authority, of reason. And therefore, Dr. Hodge, in a day of risk, has actually taken away the shelters of the truth, and made the Church weak just in the spot of her worst necessity.

Witness how these points in the truth's enclosure are seized upon by these men by a sort of natural instinct. Not only do they deride authority the moment it is divorced from the support of reason; not only do they declare (and justly too) that it is then utterly incapable for either testimony or miracle; but they seize upon some of the ten points we have given, and turn them, as it is sad to see, against the belied religion.

Mill says :—" If instead of the 'glad tidings' that there exists a Being in whom all the excellences which the highest human mind can conceive exist in a degree inconceivable to us, I am informed that the world is ruled by a being whose attributes are infinite, but what they are we cannot learn, nor what are the principles of his government, except that 'the highest human morality which we are capable of conceiving' does not sanction them; convince me of it, and I will bear my fate as I may. But when I am told that I must believe this, and, at the same time, call this being by the names which express and affirm the highest human morality, I say, in plain terms, that I will not. Whatever power such a being may have over me, there is one thing which he shall not do: he shall not compel me to worship him. I will call no being good who is not what I mean when I apply that epithet to my fellow-creatures; and if such a being can sentence me to hell for not so calling him, to hell I will go" ("*Exam. of Ham.*," Boston Ed., vol. i. p. 131).

Oh, how sad the exhibition, when such a recreant as Mill puts himself more right, even for an instant, than the Church, as to the attributes of the Almighty!

And listen to Spencer, too,—"He may think it needless, as it is difficult, to conceal his repugnance to a creed which tacitly ascribes to The Unknowable a love of adulation such as would be despised in a human being. . . . There will perhaps escape from him an angry condemnation of the belief that punishment is a divine vengeance . . . and that conduct is truly good only when it is due to a faith whose openly-professed motive is other-worldliness" ("*First Principles*," pp. 120, 121).

Bad morals are like the weak joints of a rheumatic. The first sour wind that blows will instantly rack them. These men find out what is putrid in religion,— as the vulture, carrion ; and down to that spot in the landscape will be the first swoop of the foul wing. Hence the value of infidels as scavenger-birds.

A father, learning from his son that by reading Spencer his College days are unsettling him on the subject of religion, writes back to him—what? That he must ask help from a higher source, better to understand such literature? that he must read more diligently the Bible, and come with more immediate faith to the help of the Redeemer? Alas! possibly!—no doubt he may have told him all this ; but the conspicuous reply which reached others, and was talked of in the place, was that HE MUST SUBDUE HIS REASON. Now that layman in the Church has learned so to talk from the teaching of theology.

So wrought not Christ. He took care of reason, and means to appeal to it in the Day of Judgment. The daintiest web of reason could not be subtler than that with which He announced His advent. He did not come barely with authority, but exerted Himself to prove it. He chose His witnesses. He scattered them over the earth. He shot a ray of light first into Persia, and the Magi of the East were made to expect His coming. He sent them to Herod. He sent Herod to the Sanhedrim. He stirred up all classes of society. That the lowest might be reached, He heralded in a still brighter way the shepherds. We see the same handwriting at His death. As His birth was accomplished where the "whole world" was being taxed, so His rising-again, when the whole world was gathered at Jerusalem. The daintiest care was manifested with the evidence. Was there an earthquake, it was at a crowded passover. Was the vail of the Temple rent in twain, it was on that hour of the year when the finger of the Priest was about to push it aside and enter. Might there be a scandal of deceit? The Jews themselves were to prevent it. They made it certain that the rising from the dead should be attested; for they made everything sure, sealing the stone, and setting a watch. All through the Augustan world God had arranged to give the most honor to an appeal to reason, and, in fact, the whole Bible, rendered in Greek, and copied in every synagogue, was open to the inquisition of the world; and Luke distinctly tells us that one synagogue was more noble than that in Thessalonica, in that they "searched the Scriptures daily," and that not to submit to them on a mere authority-faith, but boldly, "whether those things were so" (Acts xvii. 11).

Rationalism, therefore, is not an "undue authority of reason." There cannot be such a thing. Religion appeals to all the reason we can muster up,—and more. Religion appeals to nothing else within us but our very strongest reason. And if a man fails at the day of final assize, the first arrow that will be shot will be one that shall convince his reason; the first pang under which he shall cry out, will be under a sense of his extreme irrationalness; and the strongest conviction that he shall have will be, that conscience and the testimony of the Spirit were all and every part of them in the domain of reason.

§ 5. *The Scriptures that Dr. Hodge quotes.*

First, all those Scriptures that teach that "the world by wisdom knew not God" (1 Cor. i. 21). This does not mean that the world had too much wisdom, but too little. A painter's not admiring his art does not show that reason is not in play, but a deficient reason. His reason lacks that æsthetic *nous* which is just as much its province when it possesses it, as a sense of cold, or a sense of truth, or a sense of power, or any other sense of which it may judge or reason.

Second, all those texts which speak of truth upon authority; all those texts which are preluded by a "Thus saith the Lord." Of course we accept them as duly as Dr. Hodge; in fact a little more so; for we are not so ready to reject them when they contradict our senses (i. p. 60). But we receive them on the authority of reason. We let our minds reach down and touch bottom at once, which Dr. Hodge is slow to do. And instead of being forced to ruin everything by admissions at the very last as to the *judicium contradictionis*, we remember that at the beginning, and we say, All things we receive on reason. We receive God on reason. So He teaches us to do. We receive authority on reason. So it is reasonable to do. We receive this calf-skin on reason. Nothing else could judge it:

but we infer after careful reasoning that it has the authority of God. And then if He bestows His Spirit, we receive that into reason. It is reason that He lights up with a forgotten and dishonored light, and we refuse to compare as though different things, faith and a human reason, and choose rather to speak of a reason blinded and imperfect, and a reason restored to itself by the finger by which it was first created.

Again, we have all those passages which speak of the sinner's foolishness. Paul had learned about this in the synagogue (*see Solomon* passim). When, therefore, he says, "The wisdom of this world is foolishness with God" (1 Cor. iii. 19; or when he says, "The preaching of the cross is to them that perish, foolishness" (1 Cor. i. 18); or when he says, "If any man among you seemeth to be wise in this world, let him become a fool that he may be wise" (1 Cor. iii. 18),—he means, not wise really or *semet ipso*, but wise *quoad* this fallen world; he means, not wise in the sense of rational, for even Dr. Hodge would admit that was not the case; he means, therefore, not wise at all, but foolish ; and means only that a man must find out that he is a fool, before his reason can ascend to the highest wisdom.

§ 6. *Injury of Dr. Hodge's Teaching to Dr. Hodge himself.*

Dr. Hodge spent a student's year in Germany. He found Germany, now half a century ago, all alive in its smaller and purer church against the encroachments of Neology. Just as another colleague in his earlier training was in a city oppressed by Prelacy, so Dr. Hodge received a bent through his whole polemic life against the assaults of Rationalism. In declining years he has turned to Metaphysics as more of an ally, and to Science as a thing to be understood; but all through his intermediate life he has denied that Metaphysics was a system, or that it was capable of progression under the conduct of the human mind.

High thought has revenged itself for the contempt.

And this in two ways appears as we search the old and the new in these final volumes.

In the old he is the strong theologer, broad and lucid, and most dexterous in written debate, and yet always disappointing the mind that goes deeply after the seeds of things; leaving his threads untied-up, and excusing himself by appeal to the mysteries that become the gospel; in the new, betraying the old man's work; we mean, respectfully, the hand of an aged convert; never having respected Metaphysics; lately drawn to it; never having vested much in Science; lately moved by it as endangering the Church;—and, therefore, offering it in that scrap-book form uniting badly with his older work, and giving the reasoning of the skeptic in clearer because more studied paragraphs than his own too late attempts to recover ground in the regions of their reasoning.

Hence passages all through the work of Dr. Hodge that injure it with scientific reasoners.

Take the first sentence. "In every science there are two factors; facts and ideas; or facts and the mind" (vol. i. p. 1). Or take the second, "Science is more than knowledge." Or take the fourth, "The facts of astronomy, chemistry, or history do not constitute the science of those departments of knowledge." It is such sentences that quietly wave back just as they are setting out, that very class of readers that Dr. Hodge would prefer to secure and influence.

So on the fourth page,—"There are laws of nature (forces) which are the proximate causes of natural phenomena."

On the 634th page he says,—"Confidence in human testimony is founded on a law of our nature"(!). This he gives as an answer to Hume's celebrated argument on miracles.

He says that he finds in the Old Testament that "Sheol is represented as the general receptacle or abode of departed Spirits, who were there in a state of consciousness; some in a state of misery; others in a state of happiness" (vol. iii. p. 717),—thus settling at a stroke all the question why the Old Testament has no such reference at all.

As an ethical truth he tells us, that "the guilt of sin is infinite in the sense that we can set no limits to its turpitude, or to the evil that it is adapted to produce;" and then, that "The crucifixion of Christ was the greatest crime ever committed" (vol. i. 544), not informing us which of Christ's crucifiers he alludes to, and then not saying how an evil can be infinite, and yet an instance of it the greatest one that ever appeared.

Steady assaults upon reason produce such reactions. We see it in certain mannerisms. Dr. Hodge assails opponents by an appeal to universals. For example, 'Such,' referring to the doctrine he supports, 'has been the faith of the Reformed Church in all ages.' This, when he is replying to a large body of the Reformed Church at the very moment. It is a feature of his book.

So an appeal to consciousness. He makes it till it comes up by a sort of instinct in every case.

Not only is God Himself a matter of consciousness (i. p. 191), and cause and effect a matter of consciousness (i. p. 210), and the immaterial nature of the soul a matter of consciousness (i. p. 36), but he comes at last to instances like these,— "It would be no more irrational to pronounce Homer and Newton idiots than to set down Isaiah and Paul as either impostors or fanatics. It is as certain as any self-evident truth that they were wise, good, sober-minded men" (i. p. 37). And I might choose Balaam or, if I pleased, Solomon, or, in lesser degree, Buddha or Confutzee, or even Mohammed, and might assert the same point with an approximate degree of rationalness. Again,— "It is intuitively true that sin ought to be punished" (ii. p. 539). Much of these three volumes is built *gratis* on just such assertion. And again, what follows? "They therefore know, that without a satisfaction to justice sin cannot be pardoned. This conviction lies undisturbed at the bottom of every human breast" (*ibid.*). Men's instincts lie in another direction, and Dr. Hodge himself, when pushed, takes refuge in atoning *mystery* (ii. p. 479). And yet now while on another beat he seizes this repeated support. Sacrifice is intuitive. "Without satisfaction sin cannot be pardoned. This conviction, sooner or later, never fails to reveal itself with irrepressible force to the reason and the conscience" (*ibid.*). And again, God's Providence! A denial of it " contradicts the testimony of our moral nature. The relation in which we stand to God as that relation reveals itself in our consciousness, implies that we are constantly in the presence of a God who takes cognizance of our acts, orders our circumstances, and interferes constantly for our correction or protection" (i. p. 35). All this, mark you, a man's consciousness. He is conscious even that a sparrow does not fall to the ground without our Father (*ibid.*). Again,—" men must enter that gate [and he means the gate of death] conscious that they have within them an imperishable life " (i. p. 36),—of course meaning, conscious of immortality, another great doctrine ; and then, propounding still another, he adds,—" life combined with all the elements of perdition ; " and, again,—" Is it not self-evident that immortal sinners need some one to answer the question, What must I do to be saved ? " A thinker who decries reason, and postpones it to a weird authority, lifts it right up in its highest act, viz., a decisive consciousness, and makes it the facile prop of half the doctrines of the gospel.

Three cases more :—

"It is intuitively true to all who have eyes to see, that Jesus Christ is the son of God, and that His gospel is the wisdom of God and the power of God unto salvation" (i. p. 260).

"The doctrine of *concursus*" (adopted by Augustine, and taught by Turrettin and many of his school) "contradicts the consciousness of men" (i. p. 604). And then, once more,—"Men no more need to be taught that there is a God, than they need to be taught that there is such a thing as sin" (i. p. 199).

§ 7. *Argument from Reason.*

The worn-out pivot-hole of a machinery that once moved with a quiet that scarce shook the building, is a fine emblem of the theologizing that has grown to be habitual in the instance of Dr. Hodge. We need not stand and cavil and say, Here was the pivot-hole, or, Here the true motion ought to have been kept. It is enough to prove what has happened simply to listen to the shake, and hear the factory reverberating to the changes of the stroke.

Dr. Hodge never tarries a moment on a single beat.

Scarcely has he told us that "the knowledge of God is innate" (i. p. 191), before he is spinning off upon the question, "How do we form the idea of God?"

(p. 339.) The phrase, "The will of God is the ground of moral obligation" (i. p. 405) seems to throw him to the other extremity of thought,—"He cannot make right wrong or wrong right" (p. 406). These are the greatest defects that can occur in any system. And really, the argument from reason in the present instance is easiest made by incorporating in it another section, viz.,—

§ 8. *Dr. Hodge's Contradiction.*

Let us be exact.

In theologizing, the ultimate appeal is to authority, or to reason.

Dr. Hodge says, To authority.

Now we might go into that direct, and show what was bewildering his mind.

In Germany men were rejecting the Trinity because they did not understand it. Did that make it less amenable to reason? In Germany men have rejected gravitation because they did not understand it. We might throw the two facts together, and show that reason is equally in place ; that she never abdicates her seat, and that all the *dicta* she has made are precisely thus, viz., the acceptance of what she never understood.

But Dr. Hodge saves us this thought. He sets up the error. Faith's appeal is to authority (iii. p. 63). He makes it impossible to misunderstand him ;—Even consciousness ultimately appeals for its veracity to faith (iii. p. 48) ; and when that thought has been made as naked as it can, he topples it over at a stroke, and tells us the *judicium contradictionis* is at last with reason (i. p. 52).

§ 9. *The Whole System Fetich.*

Why does not Dr. Hodge sail under the Duke of Norfolk?

He may say, He does not believe in Alacoque.

But why not?

He may say, It is all a farce, and disgusting in the Nineteenth Century.

But why? It is done by authority. The appeal is twofold, first to God, and then to His greatest Church, a Church so real that Dr. Hodge would give to it deeds of lands in the protoplasm of our teeming West (*see Letter to R. R. Directors*).

Why not follow it into France?

Dr. Hodge will say,—The very proposition is absurd.

And we see at a glance that authority is nothing at all till we know what authority it really is. The Bible is boiled and pressed-out rags; God is Buddha or Olympic Jove; authority is an old wives' fable,—till reason has taken it in hand, and told which Bible to take, and which God is the true Almighty. We are not even bound to take any, till reason has told us that fact, and been the foundation of our whole religiousness.

This life spent in assaulting reason has tinged the Church with a dangerous superstition. Our cause should climb out of it. Beginning with a lapse of holiness we have let reason go, as in the

Old World picture (Rom. i. 28). Possessing the learning of Egypt, let us not possess her cats and crocodiles. Let us remember that religion is of God, but that it is also for and by man; that it inhabits all his reason; that it appeals to all his conscience; that it inspires all his affection; and that these three things together are not man in a jangling trinity, but man in different lights, reason and conscience and affection being the one mind as a revelation of the Most High.

The testimony of the Spirit we hold as he does, but does he not immediately tell us that it is not visible save in its effects? That Spirit restores our reason. That reason is our inward spirit. That spirit is all our nature. It is not to be divided into halves, or pitted against faith or piety. It is that which *is* pious, and that which *has* faith. And the man who "subdues his reason" may start for France to-morrow; he is the Greek, believing Priaps; he is the Latin, thumb-screwing Galileo; he is bedraggling what he ought to pick up, and bedrugging the whole of himself to get himself ready to get religion.

BOOK VI.

FETICH IN ORDER.

CHAPTER I.

AS RITUALISM SHRIVELS DOCTRINE, SO DOCTRINALISM SHRIVELS THE CHURCH.

WHEN a Church holds up holiness as the highest end of God and as the chief object of all His people, it keeps rite and doctrine both in place. It is the juice that animates the tree. But when, for any cause, holiness steals away, rite and doctrine both become a shell; and the Church, as for historical causes it has become attached to either, thickens that one,—lays an accent upon it, so that rite or doctrine becomes its superstitious means of arriving among the blest. Either, thus exaggerated, shrivels and neglects the other. The Papists, the great Ritualists, shrivelled doctrine so that Luther had to expand it afresh. And we, if we are advanced as Doctrinalists, have shrivelled rite. There are signs of this in the modern Church. We talk of societies when we mean organized Churches. We talk of a demission of the ministry, when it is the solemn office that Christ instituted among men (see Princeton Rev., 1859, p. 369). We plan a rotation of the eldership. We seize upon an American creed,* and upon elders as the "representatives of the people." Every thing that makes the Church an organized authoritative court, representing Christ on earth, put in being by His own hand, and made successive by directions of His lips, it is the tendency of modern Protestantism to ignore. Given a healthy Church, instituted by Christ, defined in external office, assured by solemn promise, built upon Peter, that is, avoiding the absurdities of the Pope, built up of living men; as Paul expresses it, upon the foundation of the apostles and prophets; and as Christ times it, the first, Simon which is called Peter, giving honor to that favored disciple as being the first to be laid in the building,—Ritualism and Doctrinalism take their departure from it by different poles. Ritualism takes Peter for Christ, and Church order for grace, and Church office as the channel of salvation; and Doctrinalism is at the stark extreme. Ritualism thickens the Church, and lays the accent upon it, and makes the idol out of the external shell. Doctrinalism upsets it altogether. It has its idol at another pole. It denies the externality of the Church, I mean as vital to it (Princeton Rev. vol. xxv. p. 250). It makes light of a pattern as having been shown in the Mount; and whereas Paul warns us to " lay hands suddenly on

* Our "Form of Government" is an American production. The Westminster " Form " says nothing of " representatives of the people." Why was it discarded? See Assembly's Digest *in loco* for the History of the change, but the reasons are not given. Why did not our " Form of Government," when first written by a Synodical Committee in 1787, preserve more the ancient features?

no man," it gets rid of that risk by demissions of the work, and by rotating the called of Christ as though they were the mere " representatives of the people " (*see Dr. Hodge's Tract on Pres.*).

This is the tendency of modern times. But Dr. Hodge, as standing in the lead, and as having brought together more of these doctrinalistic traits than any other expounder of the gospel, shrivels the Church, too, more than any that have gone before him; carries his conclusions to further points; and sells his own authoritative Presbyterianism to all those busy claimants (who are glad that he is lax) of a more positive order, and a more downright institution by the Redeemer.

CHAPTER II.
THE TRUE IDEA OF THE CHURCH.

JESUS CHRIST instituted the Lord's Supper, and said, "This do in remembrance of me" (Luke xxii. 19). The word had hardly gone out of His mouth, before inspired writers began to emblematize the sacrament in every possible way. He Himself says, "Except ye eat the flesh of the Son of Man, and drink His blood, ye have no life in you" (John vi. 53). Ritualism and Doctrinalism might come in here, and carry the ordinance off in the same opposite directions. Ritualists might say, The ordinance is a saving feast; and Doctrinalists might say, It is not necessarily a feast at all.

Now, how perfect the likeness to the Church. Christ prayed all night, and then ordained apostles

(Luke vi. 12, 13). He gave them the most stringent powers (Matt. xvi. 18; John xx. 21, 23). He settled them upon the most permanent and far-reaching work (Luke xvi. 15). When cloud clears, we find them with a Septuagint name,* and with an organization plainly resembling the Old Testament type. So actual is this Church, that inspiration gives directions for handing it down (2 Tim. ii. 2); and so authoritative is it, that it girds itself with disciplinary powers (John xx. 23); and so practical as an external fact, that it plunges its life into the very midst of the most worldly service (Acts vi. 3).

And yet our Lord, just as He commands baptism, and then employs it as an emblem (Mark x. 39; see also Matt. iii. 11); and just as He commanded sacrifice, and then speaks of it as saving (Lev. x. 17; Mal. iii. 10); and just as He appointed circumcision, and allowed it to be an emblem in the Church (Rom. ii. 28, 29); nay, just as He called Abraham, and made his seed typical of actual salvation (Gal. iii. 29),—so He seizes upon the Church, and all through His Word makes it the symbol of "the body of believers." Zion (Is. lii. 1), and Canaan (Ps. xcv. 11), and Egypt (Rev. xi. 8), and the cross (Gal. vi. 14), and His blood (1 John i. 17), and His flesh (John vi. 51), and the Temple (1 Cor. iii. 16), and the Altar (Heb. xiii. 10), and the Holiest of All (Heb. x. 19),—all teach in the manner of the East. And the moment a sentence fell out

* Εκκλησια.

that used them figuratively, Ritualists gathered it up. The very congregation of God grew to be regarded as "the body of true believers" (John viii. 33); and though ritualistically, as in the Papal Church, yet with the same failure to consider that the words were employed as emblems.

Now let us press all this into our service at the present day.

The word *church** is used ninety-six times in the New Testament for an external organization. It is used twelve times, and hardly that, as baptism is when it is applied to actual salvation. Who would ever dream that these twelve cases would define the Church? Burnt sacrifices were actually prescribed, and men continued externally to heap them up. Abraham was actually called, and Jews continued outwardly to live apart. Baptism was ordered, and men offered it; and the Lord's Supper, and it is externally continued in the world. All these were used as emblems; but sober sense kept on the external rite. But the Church is instituted, and here comes up a man who says that it is "the body of true believers;" that it is not essentially external; that it is visible primarily because saints are visible; that it is not therefore obligatory in any decypherable form; and when asked for his proof, points to the very expression, the Church the body of Christ (Col. i. 18), just as a man could deny that baptism is external because it says, "baptism doth now save us" (1 Pet. iii. 21).

* Εκκλησια.

The doctrine of the Church is exactly what good men would read if it had never been employed in emblems; external, like the Septuagint norm; actually framed by Christ; actually formed with powers; actually braced with office; intended to be continued by descent; and armed with salutary rules, and with becoming authority, under the will of the Redeemer. If any one says, Is there but one pattern therefore? We say, Unquestionably but one. If he asks, What, therefore, that pattern is, We ask him, What is the pattern for baptism? Most men thus far believe that that is an external rite, and that it is to be administered in a certain way. Neither baptism nor the Lord's Supper have come to be doctrinally impugned. But yet as to their specific forms there is some obscurity. Who doubts that we should examine our very best? and who hesitates that out of our most careful search there should emerge that rule for the two that shall show our most docile purpose to follow our Deliverer? Presbytery is better than Prelacy. So think certain Christians. Prelacy is better than Presbytery. So think certain others. One is undoubtedly wrong. But how much more wrong the man that says that the Church is not in its own essence external; and how thoroughly would Baptism be destroyed if it were treated with the same use of the figures of the East.

CHAPTER III.

DR. HODGE'S IDEA OF THE CHURCH.

DR. HODGE's idea is that the Church is "the body of true believers" (*Idea of the Church*, Princeton Rev., vol. xxv. p. 251).

His reasoning is very peculiar. He calls attention to the fact of the acceptance of the Apostles' Creed (*ibid.* p. 249). He then quotes from the Creed :—" The Holy Catholic Church ; the communion of saints," and inaugurates his plan at once on the basis of this grandest Symbol. But let us attempt a similar. The Church is the Holy Ghost ; for listen again (out of the Creed),—" The Holy Ghost ; the holy catholic church." Or once more,—The communion of saints is the forgiveness of sins ; for listen again,—" I believe in the Holy Ghost ; the holy catholic church ; the communion of saints ; the forgiveness of sins ; the resurrection of the body ; and the life everlasting." Ought we not to be exempted from such arguments ? Or, turning the proof into just the opposite, does this Creed ever repeat ? Is not " the communion of saints " *ex intentu* a separate count ? And is not our " Confession " right when it makes a chapter on the " Holy Catholic Church," and follows it immediately after with one on the communion of believers ?

Dr. Hodge, however, with this impulse from the

"Creed," goes on to expound his system. Our Confession, with one sentence at the first to guard us against the emblematic sense, gives five sentences out of six to its working term, viz., "the visible church, which is also catholic or universal." Dr. Hodge never quotes from it. He quotes other Confessions; from the Helvetic; from the Augsburg; from the Nicene; from the Belgic; and from the Lutheran more than once; but he never notices his own Westminster symbol. Indeed he differs from it; not simply in the major case, but in another distinct averment. The visible church he argues cannot be catholic; and yet here we have the distinct language of the Divines, "the visible church which is also catholic" (Conf. chap. xxv. 2).

But as more fair let us give the whole chapter which stands as our Confession:—

Chap. xxv. "*Of the Church.* I. The catholic or universal church, which is invisible, consists of the whole number of the elect that have been, are, or shall be, gathered into one under Christ the head thereof; and is the spouse, the body, the fulness of Him that filleth all in all.

II. "The visible church, which is also catholic or universal under the gospel (not confined to one nation, as before under the law), consists of all those throughout the world that profess the true religion, together with their children; and is the kingdom of the Lord Jesus Christ, the house and family of God; out of which there is no ordinary possibility of salvation.

III. "Unto this catholic visible church Christ hath given the ministry, oracles and ordinances of God, for the gathering and perfecting of the saints in this life to the end of the world; and doth by His own presence and Spirit, according to His promise, make them effectual thereunto.

IV. "This catholic church hath been sometimes more, sometimes less, visible. And particular churches, which are members thereof, are more or less pure, according as the gospel is taught and embraced, ordinances administered, and public worship performed, more or less purely in them.

V. "The purest churches under heaven are subject both to mixture and error; and some have so degenerated as to become no churches of Christ, but synagogues of Satan. Nevertheless there shall be always a church on earth to worship God according to His will.

VI. "There is no other head of the church but the Lord Jesus Christ; nor can the Pope of Rome in any sense be head thereof; but is that antichrist, that man of sin, and son of perdition, that exalteth himself in the church against Christ and all that is called God."

On the contrary, Dr. Hodge says, "In that symbol of faith adopted by the whole Christian world, commonly called the Apostles' Creed, the Church is declared to be 'the Communion of Saints' . . . It is obvious that the Church, considered as the communion of saints, does not necessarily include the idea of a visible society organized under one definite form. . . . It is not presented as a visible organization to which the form is essential" (Princeton Rev., vol. xxv. p. 249). Again, "It does not include the idea of any external organization. . . . . The Church, therefore, ... is not necessarily a visible society. . . . . It may be proper that such union should exist: it may be true that it has always existed; but it is not necessary. The Church, as such, is not a visible society. All visible union, all external organization, may cease; and yet, so long as there are saints who have communion, the Church exists, if the church is the communion of saints" (*ibid.* p. 250).

Let us paraphrase this. 'The A. B. creed says, "Baptism is with the Holy Ghost and with fire." Baptism, therefore, is not essentially external. It is not necessarily a visible rite. It may be proper that such rite be shown forth, but it is not necessary. Baptism, as such, is not a visible administration. Such usage may cease; but, if the Holy Ghost continues to be administered, the man is baptized—if baptism is by the Holy Ghost and by fire.'

Or, take another case. 'It was promised to the seed of Abraham that they should inherit the earth. The Jews took a carnal view of this; and supposed that external relationship would secure salvation. Therefore, the *ecclesia* of the ancient seed was not in its essence external; and the gathering of the sons was not an essential fact in the congregation of the Lord.'

"So far, therefore, is the Apostles' Creed from

representing the Church as a monarchy, an aristocracy, or a democracy; so far is it from setting forth the Church as a visible society of one specific form, that it does not present it under the idea of an external society at all. The saints may exist; they may have communion; the Church may continue under any external organization,* or without any visible organization whatever" (Princeton Rev., vol. xxv. p. 250).

"By this statement it is not meant that the word *church* is not properly used in various senses. The object of inquiry is not the usage of the word, but the true idea of a thing; not how the word church is employed, but what the Church itself is; who compose the Church?" This is very important; otherwise, we might be confused by the fact that an invisible church is spoken of under a metaphoric head. Dr. Hodge properly expounds, that we are speaking of the working title, that unmetaphoric sense that we drop-to when we treat of baptism or the Lord's Supper. "What is essential to the existence of that body to which the attributes, the promises, the prerogatives, of the Church belong?" (*ibid.* p. 252.)

"The Church of God," Dr. Hodge replies, "is the whole number of the elect; the Church of Corinth is the whole number of the called in that city" (*ibid.* p. 255). The Church of Princeton, therefore, is the pious people there, even though they contumaciously refuse to unite in any communion.

* The Young Men's Christian Association, for example.

"The descriptions of the Church given in the Word of God apply to none but true believers" (*ibid.* p. 261); as, for example, the Church in Laodicea, to whom Christ says, "I will spue thee out of my mouth" (Rev. iii. 16).

And in the late speech before the "Alliance," Dr. Hodge thus stands as the representative of the Presbyterian body:—"Nothing external is essential to the being of the Church" (Tribune Report).

And then, lastly:—"If ordination be analogous to an appointment to office, as a king or president appoints a man, then no man is a minister who has not been commissioned by due authority. But if, as we Protestants believe, no Church can make a minister any more than it can make a Christian [prolonged applause], then," etc. That is,—Under the same shrouding of externality and external order in the Church, not only can a "man [be] a minister who has not been commissioned by due authority," but "no Church can make a minister any more than it can make a Christian" (Tribune Report).

These views, and others of which we are yet to speak, were in the very eye of the world, and with every motive from the presence of a different belief to represent with care his own system.

CHAPTER IV.

DR. HODGE'S ARGUMENT FOR HIS IDEA OF THE CHURCH.—ITS FALSENESS.

WHEN men are advancing truth almost universally believed, they may be sometimes negligent in their reasoning, and may offer proofs either from reason or the word of God, suggestive rather than demonstrative in their real nature. The ship may sit loosely in her rigging if there be days of calm. But Dr. Hodge, in extemporizing a course on the "Government of the Church" in the Seminary at Princeton, followed a Professor who taught just the opposite system; was followed by a Professor who restored that system's books, and taught it over again; was exhorted by "Directors" to desist, written to and written of in respect to the peculiarities of his faith, and, therefore, had every reason to ascertain its imagined evidences.

Near a quarter of a century has not made that reason less; yet when that Church, through many parts of her, has quieted her fears under the supposition that that subject had gone into other hands, Dr. Hodge has not surrendered it; he has been busy multiplying his proofs; he has influenced for a score of years the ministry of the Church; he has founded a school that it may be difficult to overset; and therefore, beyond all form of doubt, he is responsible for the most solid proofs, and for the most unchallengeable scheme of arguing, if he is to uphold his system.

What is that scheme?

We can present it fairly to be tried if we discuss it under the seven heads under which he offers it to our consideration.

Dr. Hodge's Church, therefore, is "the body of true believers," and his proofs are: First, "the word *church;*" second, the equivalent expressions; third, the descriptive terms; fourth, the "attributes ascribed" to it; fifth, "the promises" made to it; sixth, the doctrine that belongs to it; and seventh, "the theory" that has been held by it, or, as Dr. Hodge expresses it, "the testimony of the Church," that it has given in respect to its own nature.

I. Now in regard to the name, that name is derived from two Greek words that mean to *call out* (εκκλησια). Dr. Hodge says, "Every εκκλησια is composed of the κλητοι of those called out and assembled" (Princeton Rev., vol. xxv. p. 256). And here at the outset let us remember that the word εκκλησια came down to Paul from the Old Testament worship. It is a bad argument that will not apply where the word was first put in use. In the Septuagint we read, "In the midst of the congregation (εκκλησιας) will I praise thee" (Ps. xxii. 22, *quoted* Heb. ii. 12; see also I Kings viii. 14; Deut. xviii. 16). Now, unless the "congregation" to which this name was given was not in any essential way external; nay, more flatly, unless the town-meeting (εκκλησια, Acts xix. 32, 39, 41) that wanted to stone Paul were the εκλεκτοι, (for it is on the term itself that Dr. Hodge builds the argument), then we can dismiss this proof at once. The term became Scriptural at Alexandria among Ptolemy's Seventy. Why they used it, and where they got it, we might most easily tell. It derived itself in a way extra-religious. It can be found in Donnegan like any other Greek. And to build a tenet upon it, and that tenet painful to the Church, is an act which the very coolest men, when they begin to challenge it, must feel the impulse to press to the actual point either of explanation or surrender.

But suppose we pass it, Suppose the word had been invented by Christ. Suppose it had been applied to nothing but the New Testament church. Suppose it were linked with κλητοι, and to be expounded by κλησις and every linguistic mate. What would that prove? Dr. Hodge says, "None but those who truly repent and believe are ever called κλητοι" (Princeton Rev., vol. xxv. p. 256). Now how unfortunate this is! How unfortunate all such universals are unless most carefully framed. The word καλεω is used with wonderful vagrancy. It has no set fitting whatever (Is. xlv. 4; Heb. v. 4), and this very articulate part of it is used for those not εκλεκτοι, and this by the way of a most distinctive caution; for we are told,—" Many be called (κλητοι), but few chosen" (Matt. xx. 16).

Passing by the fact, therefore, that it is an Athenian term (εκκλησια), and passing by what follows, viz., that it cannot found an argument, and omitting all complaint that this surd evidence succeeds that like one from the Creed, ("The Holy Catholic Church;

the Communion of Saints,") we might admit that Christ framed the word, and yet hold that it is the best chosen Greek exactly to express the relation of an external church as called out of an ungodly nation (see Jud. xx. 1, ἐξεκκλησιασθη).

II. But let us go on to the next argument.

He says, "Those epistles in the New Testament which are addressed to churches, are addressed to believers, saints, the children of God. These latter terms, therefore, are equivalent to the former. The conclusion to be drawn from this fact is, that the Church consists of believers" (*ibid.* p. 258). Now let the reader hunt this up in the pages of the Princeton Review, and cast his eye one paragraph before, and read,—" All the professors in Corinth are called saints, sanctified in Christ Jesus, the saved, the children of God, the faithful, believers, etc., etc. . . . Their being called believers does not prove that they were all believers" (*ibid.*). And yet the very argument now is, that the churches being addressed as believers, proves this very thing, viz., that they are the body of Christ; so punctual is Dr. Hodge in dropping something that shall contradict him in the close neighborhood of his extremer teachings.

His is our very most lucid answer.

"The Church is the body of professed believers with their children" (Conf. of Faith, chap. xxv. 2). Our confession gives us our working sense. But the body of professed believers are to be treated hopefully. So must always be the case. Paul cries,—"We accept it always, most noble Festus." He treats him hopefully. We cry,—"We humble ourselves before thee, O God; we love thee and express our gratitude." We treat ourselves hopefully. It is the style, beyond a doubt. We call all people saints who are at the communion table. Moreover, holiness is one mark of the Church. It belongs to its externality. If a church ceased to be holy, it would cease to be one. That is, if externality is essential to the church, so is it that it should be holy. An attempt at the New Testament frame is not so vital to the body as that it should have some members, at least, who actually have turned from their wickedness.

But how precisely can Dr. Hodge make his point? He says (Princeton Rev., vol. xxv. p. 258), —" All the professors in Corinth are called saints." Then, of course, if the Church is called the saints, he has shattered his own argument. The beloved of God, the called of Jesus, the sanctified of Christ, the "elect unto obedience," or any other excellent people, may be all addressed under the title of the Church, and Dr. Hodge has himself destroyed the possibilities of the needful demonstration.

III. So now the third argument.

(1) "The church is . . . the body of Christ," Eph. i. 22 (p. 262).

Dr. Hodge may make his election, either to count this one of those strong cases where professors of religion were talked to as though they were real Christians, or do what we prefer, couple it with

Old Testament emblems. Zion is spoken of as a bride (Is. lii. 1); Moriah as God's holy hill (Ps. xliii. 3); David as the King of Saints (Ps. cxxxii. 17); and Joshua as Jesus Christ (Zach. iii. 1); and yet, who doubts that there were such external facts? and who, on the ground of the like, ought to deny the reality of Christ's organized kingdom?

The arguments, therefore, are like Fairbank's scale. They have but a single tread. Once know that the church has an emblematic sense, and why multiply the refutation? (2) "The church is the temple of God" (p. 263); (3) the church is the family of God (*ibid.*); (4) "the church is the flock of Christ" (*ibid.*); (5) "the church is the bride of Christ" (*ibid.*); "living stones" (p. 264); "elect, precious" (*ibid.*); "purchased by His blood" (*ibid.*); pages of this Review all hang upon a single fact which our Confession articulately expounds. There is an emblematic "Bride," bearing the same relation to the working Church that the Spouse did to the external Zion. The thought of her debauched the Jews, just as the same fancy misled the Papists; and in setting that right, viz., what Israel trusted to the blood of Abraham, and what the Papists trust to the external church, we would destroy her externality altogether, that is dethrone the fact because Idolatry has abused the emblem.

IV. We advance to the weaker arguments of Dr. Hodge when we strike his fourth point, viz., the three marks of a church,—holiness, unity and catholicity. Who framed those tests? If the Papists, of course they suit their hierarchy; and if Dr. Hodge, of course they apply to the "body of true believers." Where is the bearing of such a proof? Dr. Hodge says, a visible church cannot be catholic (*ibid.* pp. 276, 278). His own "Confession" says that it can and is (*Conf.* xxv. 2). Who shall decide? And furthermore, what if it is not? who shall condemn? We believe that the visible church is one and catholic and holy, but if it is less perfectly so than the body of Christ, and if they do not usefully apply, there is not the slenderest harm had we to dismiss the three marks as all impertinencies.

V. And so of the next argument:—The church has promises (p. 279); first, of the Divine Spirit; second, of divine teaching (p. 280); third, of divine protection (p. 281); fourth, of divine success (*ibid.*); fifth, of holiness and salvation (p. 282.) The visible church is promised these things as her general gift. She is to extend from sea to sea; she is never finally to cease; the gates of hell are never to prevail against her (Matt. xvi. 18). And if Dr. Hodge means anything more specific than this, he is mixing the accounts of what is visible and the Bride of the Redeemer.

VI. The sixth argument is what Dr. Hodge styles "doctrinal." We have treated it virtually under the second chapter.* It is not necessary to choose between Ritualism and Nothingarianism. There may be a valid *Ecclesia* inside of the Pope and outside of a mere mystic spiritism.

* On "*the True Idea of the Church.*"

VII. We hurry on to the last point,—Dr. Hodge's testimony of creeds.

We have seen how he omits his own creed, and what happy testimony that creed bears to the externality of the Church.

We admit that those he quotes incline too much to the invisible. And why? They were wrestling with the idolatries of Rome. They were forms starting back from the loathsome embraces of the sorceress. The Second Helvetic seizes upon the Creed, and uses it as Dr. Hodge does:—"The Holy Catholic Church, the Communion of Saints." (chap. xvii.). But still they are behind Dr. Hodge. The Augsburg says, "The church is a congregation of saints" (§ vii.). So say we. A Free Mason Lodge is a congregation of philanthropists. So they professedly are. And yet they may not be philanthropists, and it still be a Lodge; and so the Confession goes on to say, "hypocrites and wicked persons are included, . . . although the church is properly a congregation of saints and true believers" (§ viii.). "In which the gospel is taught,"—that is the next expression,—"and the sacraments rightly administered;" that looks much like externality. And in searching through all the Creeds, not simply as quoted by Dr. Hodge himself, but in any part of their Confessions, we find not a single word like this,—an avowal that nothing visible is essential to the church (Princeton Rev., vol. xxv. p. 249), and no specific form intended for her order (*ibid* p. 250).

CHAPTER V.
PRACTICAL MISCHIEFS OF DR. HODGE'S IDEA OF THE CHURCH.

No one who owns an estate, or who has a note to collect, or an account that he wishes to turn into money, but must value to the last degree the authority of an outward government. No one who has a soul to save, but must value for himself and his children the authority of Christ. I need not unfold the reasons. That there be discipline, and that there be power to enforce it somewhere, Presbyterians will see at a glance, and accept without the necessity of argument.

Christ's authority is of two sorts, direct and indirect; as it is direct, He is the Lord of the conscience; as it is indirect, He has committed it to the Church; Christ's direct authority Dr. Hodge does not meddle with; that which is indirect he practically overturns; for it is amazing how vast the difference between Dr. Hodge's account and the account of Christ, for example, to his servant Peter (Matt. xvi. 18).

Now, forbearing all dispute; taking with us the judgment of the mass on ecclesiastical control; knowing the value of it; believing where we would be, without it; and not staying to decide whether the gift of the keys (Matt. xvi. 19), and the binding on earth (Matt. xviii. 18), and the remitting of sin (John xx. 23), imply, whether specifically or not, an administered outward government; allowing all these discussions to take care of themselves; there is enough of formed belief established in the education of our people to reveal to them that the indirect authority of Christ is of the last importance to the welfare of believers.

But now how would this authority be the best? —if Christ formed a government and gave it actual rights? or, if He let things swing in the loose way in which Dr. Hodge has depicted? He called Abram, and made him an actual State. He called Moses, and made his an actual decree. He called Levi, and made him an actual hierarchy in the Church. And so, if He called the Twelve, and built upon them an actual commonwealth and court, whether would that be better, or a church not necessarily external, if He desired practical authority and wholesome government in His kingdom in the world?

Now, so easy is a true reply, that the plan that Dr. Hodge takes sells us to all manner of error. If the Church has no definite form I can indulge my tastes about it. Dr. Hodge half admits this. Again, if the church is not essentially external (Princeton Rev., vol. xxv. p. 249), I can do as I please about it. Men will often not join the church. Among vulgar saints I will flee to one that is refined. I will indulge my intellect. If I can get business-propping, I will be moved by that. Who among the sons of clay will stay lower in society, or do worse as to estate, if they dare do as they desire? Because, mark it:—The Church is the body of Christ, and no outward form is essential to the house of the Redeemer (*ibid.*).

Nay, suppose I have scruples just there. Suppose I have begun to doubt. Suppose I have become disgusted with this unpositive idea. Suppose I am moved against it by a church that holds the very reverse. What will be the result? In the vast majority of cases, a desertion to the side that is more Scriptural. It may be half Papal in all other respects, but its show of insistence here will absorb many a saint who has been trifled with with this theory of the kingdom.

And how sad it all seems! For now is a time for a more than usual reach of Presbyterian authority.

Give Dr. Hodge his way, and we bind it hand and foot just at the time when it ought to be more gravely positive. The Pontificate failing in the East; the influence of the Latin Church failing conspicuously on the part of the Latin peoples; the power of the Romish creed seating itself in a higher race, and in England, even in the most intellectual seats, nursing itself among the very highest of the people; when it is alarming us by its accumulation of estate, and taking us by surprise by the advance towards it of the chief church in all English-speaking Christendom, it is a bold feat of Satan just at that hour to tie the arm and emasculate the strength of just that body that should be most on the alert, and is certainly best able to resist the evil.

A Romanist knows Dr. Hodge's theory to be false. He can easily show it to the very feeblest of our people. Let it become the plan that is seated in the doctrine of our Seminaries, and the fight is over. So far as divine polity determines the polemic of his faith, the game is on his side. Dr. Hodge has rested our case upon a hold that can be never tenable.

CHAPTER VI.

DOCTRINAL MISCHIEF OF DR. HODGE'S IDEA OF THE CHURCH.

TRUE to a working at which we have already hinted (p. 255) a sacrifice of order to doctrine reacts into a sacrifice of the doctrine itself. It is well enough to say that the water floats the ship; we must not forget that the canal holds the water. The apostle explains it;—"The church which is the pillar and ground of the truth" (1 Tim. iii. 15). Touch authority, therefore, and what is the result? First, error in the church itself. She is not able to expel it. Give Dr. Hodge his wish, and the very faith of which he would make so much, is unguarded in its adoption among its ministers.

Again, it is unguarded in the world. The Church, which is its natural defence, cannot enforce it. If it do, the man flies. If the church has no external essence, the man leaves it at the breath of his lightest desires. We, therefore, build up other churches. And as those churches have not our faith, we build up other faiths. The no-fence husbandry loses its flock to where the fences are kept up. And so in two ways, first by indulgence among ourselves, and second by losses into the fields beyond, we build up error. We make everything of doctrine at the first, and nothing of doctrine as the final consequence.

CHAPTER VII.

ON THE QUESTION, WHAT IS A TRUE CHURCH OF GOD?

To the argument that all Protestant churches which are characterized by piety, are true churches of Christ, and, as they differ in external form, that therefore no specific form, and, as a consequence, no form at all, is essential to the idea of a church, we reply by saying that there is no such question as,—What is a true Church of Jesus Christ?

Books have been written about it, but they may as well have been written on the question, Where does the red ray end and the orange ray begin in the common spectrum? Indeed there are multitudes of questions, and, of course, multitudes of meanings in the inquiry,—What is a true church of Christ? In one sense it means, What is the church that Christ organized? What is that form of government which He left, just as He left a form of Baptism and a form for the administration of the Holy Eucharist? In this sense there is but one church. In this sense I fear there is no church. In this sense each particular organization arrogates its own. And all the marks of the church; its unity, for there is but one such church; its catholicity, for all should belong to it; its holiness, for it is no true church if it has no holiness at all;—these marks, which are not Scriptural, which are not definite, which are of no practical authority, and which belong to no question at all in any usual sense, fit or do not fit, just as it may happen that we have this question or that under the form of enquiry that we may be pressing at the time.

A *true Christian*,—there there is a different part of speech. There there are marks and limits. There there is a settled boundary. A man is either lost or saved ; and there it is at each conscious instant. But the moment one point is settled, viz., What organism did Christ found on the earth? then all splits up into parts. It is no more one broad inquiry,—What is a true church of Christ? but a question of different degrees ; nay, a question of different acts. It is really asking,—With whom may I hold communion? and when that becomes the point, it is settled in different ways.

Hence really there are different boundaries : for ministers, a very narrow one. Is the Methodist a true church of Christ? I say, No. We will not accept her Councils. But you meet me in a different mood and I answer,—Yes. It all depends on what you want to know. Is she a church from which I can take sermons ? Yes, certainly. From which I can take baptisms ? Unquestionably I can. From which I can receive members? Undoubtedly. What do you mean then ? Of course you mean to ask all these questions in a single one. But as unfortunately they each admit of a distinct reply, your question is not a good one. There is no true Church of Christ. His body might be so described, or the Papacy might be so imagined, or my

Presbytery might be so believed, but under the working definition of the word that question, as usually pronounced, is utterly misleading as to its full solution.

And why do we need it? I ask, Is this the City of London? A stranger tells me, Yes. A listener instantly denies it, and tells me, No. What do they all mean? They plunge into a heated quarrel. What is the difficulty? The difficulty is that I asked what is not a question. Am I out in the country, or am I yet got into the town? Here are the gas lights, but then not yet the municipal control. What did I mean? Besides this is not London at all. It would be Westminster, even if I had got into the limits. Which all means that there is no such City as London; that is, that I can quarrel about it to any conceivable length : that is, that if I am in a practicable frame, I can utter many a sentence about it, and talk of the City of London, and be well understood; but when I begin a polemic the language immediately fails. I must drop these broad terms, for they are never technical. I must settle each question by itself. Is the Quaker a true church? No. I will not settle its ministers. Nevertheless, is it a true church? Yes, I will gladly listen to their preaching. Why? is it a true church? No, for it does not respect the sacraments. Yet it must be a true church, for what singular piety in its communion! Thus we battle the watch, though we find that the Quakers themselves believe in no organized church, and do not contend that theirs is an εκκλησια at all!

The church of God is precisely like the ordinance of baptism; we are bound to administer it. We are bound to find out its form. If it varies a little, we may or may not accept it. It is a matter of judgment. There is a true church, and there is a true baptism. They have been instituted by Christ. Baptism is external, and is by sprinkling. The church is external, and is presbytery. This is my firm belief. Is any other church true? No. Is any other church true? Yes. It depends on what you mean. You have propounded questions that are not framed so in the language of holy writ. And if you go all the distance of Dr. Hodge, and tell me, The Church is the body of Christ, I reply that baptism is the baptism of the Spirit; but if baptism thus answering as a trope does not thereby cease to be a ceremony, so the church must be bodily set up; and to be thus set up, it must possess a structural form, and that form must be obligatory upon men like the form of baptism.

THE END.

www.ingramcontent.com/pod-product-compliance
Lightning Source LLC
Chambersburg PA
CBHW032221230426
43666CB00033B/465